Advanced Structured Materials

Volume 78

More information about this series at http://www.springer.com/series/8611

Bhagwanjee Jha · Devendra Narain Singh

Fly Ash Zeolites

Innovations, Applications, and Directions

Springer

Bhagwanjee Jha
Department of Civil Engineering
Dr. B. B. A. Government Polytechnic
Karad (D.P.), Dadra and Nagar Haveli
India

Devendra Narain Singh
Department of Civil Engineering
Indian Institute of Technology Bombay
Mumbai, Maharashtra
India

ISSN 1869-8433 ISSN 1869-8441 (electronic)
Advanced Structured Materials
ISBN 978-981-10-1402-4 ISBN 978-981-10-1404-8 (eBook)
DOI 10.1007/978-981-10-1404-8

Library of Congress Control Number: 2016939940

Printed on acid-free paper

This Springer imprint is published by Springer Nature
The registered company is Springer Science+Business Media Singapore Pte Ltd.

Preface

It gives us great pleasure in presenting this book, which mainly focuses on the latest trends in the field of utilization of fly ash. Though various conventional applications of fly ash in the construction sector are very well known to the society, a considerable amount of the fly ash is still lying unused in most of the thermal power plants and hence is a potential threat to both human and environmental health. In this situation, there is a need to focus on the innovative ideas for the value addition of the fly ash and enhance its utilization.

In this context, a thorough review of state-of-the-art on the fly ash utilization and various innovative efforts made by the researchers, in the recent past, for synthesis of zeolites (popularly known as fly ash zeolites) from the fly ash are presented in this handbook. These zeolites are being used for various industrial applications such as catalysts and adsorbents for the contaminated soils and sludge. In addition, a novel methodology for purification of some low grade fly ash zeolites is presented in this book. Attempts have also been made to provide a single source of the literature on the fly ash and its recycled products (the polycrystalline fly ash zeolites) as an easy and quick reference for the researchers and industrialists.

The book is organized in the following eight chapters.

Chapter 1: Introduction
This chapter has been written to summarize the main background of writing this book.

Chapter 2: Basics of Zeolites
This chapter is about the basics of zeolites. In addition, attempts have been made to discuss about the parameters of the fly ash, which are quite crucial for synthesis of the fly ash zeolites by chemical activation method.

Chapter 3: Conventional Methods for Synthesis of Fly Ash Zeolites
This chapter showcases available methodologies for synthesis of the fly ash zeolites and some critical issues, associated with them, which need further attention of the researchers.

Chapter 4: Mechanism for Zeolitization of Fly Ash
This chapter logically presents the mechanism of chemical transformation of the fly ash into zeolites. Attempts have been made to highlight the effects of alkali activation on the overall characteristics of the fly ash. Also, a need to innovate novel methods of synthesis of the fly ash zeolites, and to improve their overall characteristics, has been the main focus of this chapter. In addition, the chapter presents details of the applications of various advanced characterization tools (viz., physico-chemical, mineralogical and morphological) for exploring the overall properties of the zeolites.

Chapter 5: Novel Technique for Synthesis and Characterization of Fly Ash Zeolites
This chapter presents innovative approach adopted by the authors for synthesis and characterization of fly ash zeolites. Also, attempts have been made to explain methods of characterization of the fly ash and the end products obtained from it. This chapter also presents a discussion on the suitability of the fly ash for its effective zeolitization.

Chapter 6: Major Findings of the Three-Step Activation Technique
This chapter deals with the inferences derived from the novel method "three step activation" of the hopper ash, which has been ascertained to be the superior ash over the lagoon ash, as described in Chapter-5, by following hydrothermal activation method. Furthermore, this chapter also showcases the outcome of the three-step activation of the fly ash by fusion method to synthesize high grade zeolite-X.

Chapter 7: Application of Fly Ash Zeolites: Case Studies
This chapter presents various applications of the fly ash and its zeolites.

Chapter 8: Path Ahead
This chapter highlights the scope of further research on the fly ash, fly ash zeolites, and the reaction by-products obtained as a result of chemical activation of the fly ash with NaOH in a hydrothermal system.

It would be our pleasure to receive critical suggestions, if any, from the readers for improving this handbook.

Acknowledgements

With immense pleasure authors would like to express their sincere gratitude to each and every one who has been associated in this endeavour viz., our family members, research scholars and the Editorial team of Springer. Needless to say, without prompt help and support that we received from our colleagues at Sophisticated Analytical Instrument Facility, SAIF, Civil Engineering Department, Earth Sciences, Metallurgical Engineering & Materials Science and Chemical Engineering Department, IIT Bombay, this project would never have seen the light of the day.

Bhagwanjee Jha
Devendra Narain Singh

Contents

About the Authors

Dr. Bhagwanjee Jha is a lecturer at Department of Civil Engineering, Babasaheb Bhimrao Ambedkar Government Polytechnic, Karad (D.P.), Dadra and Nagar Haveli (India). He obtained his Ph.D. from the Indian Institute of Technology Bombay, India. His academic and research experience on fly ash spans more than two decades. His research is multidisciplinary and focused on the development of environmental cleanup strategies by applying fly ash zeolites. He is the recipient of prestigious "*Pedagogical Innovation Award-2015*," instituted by Gujarat Technological University, Ahmedabad, Gujarat, India. E-mail: jha66b@yahoo.com

Dr. Devendra Narain Singh is an institute chair professor at the Department of Civil Engineering, Indian Institute of Technology Bombay, India. He obtained his Ph.D. from the Indian Institute of Technology Kanpur, India. His major area of research and practice is *environmental geotechnology*. His research is multidisciplinary and focused on geomaterial characterization. He is the editor in chief of "Environmental Geotechnics." He is a fellow of American Society of Civil Engineers (ASCE) and Indian National Academy of Engineers (INAE). E-mail: dns@civil.iitb.ac.in

List of Figures

List of Tables

Abbreviations

AAF	Alkali activated fly ash
AAS	Atomic absorption spectroscopy
AHA	Activated hopper ash
ALA	Activated lagoon ash
BET	Brunauer–Emmett–Teller
BJH	Barrett–Joyner–Halenda
CEC	Cation exchange capacity
CFA	Coal fly ash
CLI	Clinoptilolite zeolite
cpm	Counts per million
CS	Crystallite size
CT	Cations
DTA	Differential thermal analysis
EC	Electrical conductivity
EDS	Energy-dispersive spectrometry
EGME	Ethylene glycol monoethyl ether
F1	Primary fusion
F2	Secondary fusion
F3	Tertiary fusion
FAZ	Fly ash zeolites
FCC	Fluid catalytic cracking
FEG-SEM	Field emission gun-scanning electron microscopy
FTC	Framework type code
FT-IR	Fourier transforms infrared
FWHM	Full wave half maximum
HSZ	High-silica zeolites
ICDD	International Centre for Diffraction Data
ICP-AES	Inductively coupled plasma atomic emission spectrometer
IMP	Impurity
IZA-SC	Structure Commission of the International Zeolite Association

JCPDS	Joint Committee of Powder Diffraction Standard
LSZ	Low-silica zeolites
NMR	Nuclear magnetic resonance
OFA	Original fly ash
OHA	Original hopper ash
OLA	Original lagoon ash
PBU	Primary building unit
PIP	Peak intensity of product
PISZ	Peak intensity of standard zeolite
PT	Primary treatment
R1	First recycling
R2	Second recycling
RFA	Raw fly ash
RZP	Reference zeolite powder
SAR	Silica–alumina ratio
SBU	Secondary building unit
SDMBA	Stearyl dimethyl benzyl ammonium
SEM	Scanning electron microscopy
SSA	Specific surface area
ST	Supernatant
STPP	Sodium tripolyphosphate
SZR	Significantly zeolitized residues
TGA	Thermogravimetric analysis
TSA	Three-step activation
XRD	X-ray diffraction
XRF	X-ray fluorescence

Nomenclature

a, b	Total number of tetrahedron per unit cell
A_i, A_t, A_f	Counting rates per unit volume corresponding to the initial stage, at time *t*, and at equilibrium, respectively
C	Concentration of either Si^{4+} or Al^{3+}
C_e	Equilibrium metal ion concentration (mg/l)
C_o	Initial metal ion concentration (mg/l)
C_p	Heat capacity
E	Electronegativity of a compound or atom
E_i	Intermediate electronegativity of any atom
E_p	Electronegativity of atom P
G	Specific gravity
H	Hardness
K	Temperature in kelvin
k	Thermal conductivity
k_c	Rate constant for sorption of metal ions (min^{-1})
$K_{d \cdot Cs}$	Distribution coefficient for C_s ion (cm^3/g)
L/S	Liquid-to-solid ratio
m	Mass of adsorbent
M	Molarity
Me	Alkali or alkaline-earth-metal atoms
m_o	Number of water molecules
M_w	Molecular weight in g
N	Charge on atom
n	Half of the number of oxygen atoms
p	Number of monovalent metal ions
P, Q, R	Atoms in the mineral
p_1, p_2	% of oxides in residue before or after a treatment
q	Amount of metal sorbed at time *T* (min)
q	Number of divalent metal ions
q_1, q_2	Quantities (in g) of either SiO_2 or Al_2O_3

q_c	Amount of metal sorbed at equilibrium (mg/l)
q_e	Ion absorbed per unit mass of the absorbent
R_{Cs}	Uptake percentage of C_s ion
R_p	Pore radius
SSA_{BET}	Specific surface area by BET method
T	Time
T_b	Boiling temperature in hydrothermal system
T_c	Cumulative activation time
T_r	Room temperature
u, v, z	Numbers of atoms in one mole of P, Q, R
V	Volume of aqueous phase
Vp	Pore volume
w	Water molecules per unit cell
W_a	Weight of EGME retained by the sample in g
W_s	Weight of synthesized product or fly ash in g
x, y, a, b, w	Variables
γ	% variation of major oxides
ΔE_p	Change in electronegativity undergone by each atom

Chapter 1
Introduction

Abstract Fly ash, a by-product from thermal power plant, is profusely available for several bulk applications as a civil engineering construction material and is very well known to the society around the globe since last three to four decades. Much of the fly ash is still lying unused in various thermal power plants and instigates daily various problems on the earth and the environmental. To deal with such issues, this book presents a new area of fly ash reuse/application as absorbent based on its physical, chemical, mineralogical and morphological characteristics, suitable for alkali activation of the fly ash. Several researchers have opined on such activations by conventional methods for zeolitization of the fly ash. Accordingly, this book is focused to compile all such inferences of the state-of-the-art, further challenges ahead, need of innovations for synthesis of higher grade of fly ash zeolites and their applications for environmental cleanup. Attempts have been made in this book to identify a suitable fly ash type for its effective zeolitization and to develop a novel "three step activation techniques" for this purpose with focus on reuse of any by-products. Based on various applications of the fly ash zeolites as absorbent and/or adsorbent and associated environmental safety, brief scope of the further research on fly ash has also been highlighted.

Keywords Fly ash · Applications · Innovations · Fly ash zeolites · Absorbent

1.1 General

The conventional applications of fly ash for manufacturing of bricks and building blocks, sintered aggregates, pozzolana cement, concrete, filler material for reclamation of low lying waste lands and refuse dumps, filling of mines, improvement of foundation soils, asphalt concrete, as a stabilizer of sub-grade and sub-bases in pavement construction, etc., are very well known to the society.

In addition, attempts have also been made in the recent past, to employ fly ash for some specific applications: (1) treatment of polluted water, (2) decontamination of less fertile soils for agriculture, (3) removal of heavy metal ions from aqueous

B. Jha and D.N. Singh, *Fly Ash Zeolites*, Advanced Structured Materials 78,
DOI 10.1007/978-981-10-1404-8_1

solutions (viz., domestic and industrial effluents/waste water. Thus the reuse potential of fly ash as an adsorbent is now somewhat ascertained and needs further detailed research for an enhancement with growing production of the fly ash with rising number of thermal power plants, to meet growing demand of electrical power.

However, a considerable quantity of the fly ash is still lying unused in most of the thermal power plants and is consistently posing threat to both human and environmental health.

In this situation, it is the need of the present generation researchers to have focus on innovations in the research on fly ash and its value addition and reuse.

In this context, this book has been written to present a thorough review of state-of-the-art and various innovative efforts taken in the recent past for synthesis of pure and improved grade of fly ash zeolites over those reported by previous researchers employing conventional methods. Also, attempts have been made to showcase a novel technique for synthesis of high cation exchanger (the fly ash zeolites) from fly ash and detailed characterization techniques for the products. In addition, based on previous researcher's findings, various areas of specific applications of the fly ash zeolites have been explored and compiled in a lucid way.

The fly ash zeolites (FAZ) are available in hydrated alumino-silicate mineral forms and are synthesized from fly ash by its alkali activation. Zeolites can be broadly classified as ion-exchangers, catalysts and molecular filters based on their characteristics and performance. As such, the principal advantage of using the zeolites in industries are their high cation exchange capacity (*CEC*), adsorption and catalytic capacities for various environmental clean-up projects such as adsorbents for removal of heavy metal ions and other wastes.

However, the physical properties, chemical composition and mineralogy of the fly ash largely influence its conversion to zeolites, and their resulting grade and overall characteristics. Further, the formation of one or more combination of zeolites mainly depends on the available contents of silica and alumina in amorphous glass form.

Over the years, several conventional methods such as hydrothermal, fusion prior to hydrothermal, microwave assisted hydrothermal and molten salt techniques have been developed by researchers for producing FAZ. However, one to two step reactions between the fly ash and alkali have been mostly reported and the reaction liquid by-product has been disposed of in the open environment without utilizing its chemical potentials. On the contrary, such alkaline wastes may cause environmental contaminations.

In addition, the separation of FAZ from the partially activated fly ash (the impurities) remains a bug bear to the researchers and industrialists. With all such challenges in view, this book is the first time presenter of a novel technique (three-step activation, TSA), which stresses for recycling of the fly ash-NaOH-water reactions by-products till formation of zeolitic residue. The FAZ (the final residue after third step reactions) synthesized by this technique exhibits exceptionally high cation exchange capacity, specific surface area and pore area.

1.2 Conclusions

Keeping in view the all above fact, this book is intended to provide present and future generation researchers a single compiled reading material on fly ash and its recycled product, the zeolites. The under graduate, post graduate and doctoral students would find this very lucid to understand the potential of the fly ash, an industrial by-product. Above all, the students and the researchers would get state-of-the art methodology and mechanism of alkali activation of fly ash, the quality of the zeolitic products, and techniques to further purify them by chemical activation method.

Chapter 2
Basics of Zeolites

Abstract Tectosilicates have been commonly established as zeolites, which are found in nature as well as synthesized artificially. Depending upon the type of source (read raw) materials used and the method(s) of synthesis adopted, properties of the zeolites would vary and hence their application as adsorbents could be different. Keeping this in view, an in-depth description of the zeolites, their types and properties are presented in the following.

Keywords Zeolites · Fly ash zeolites · Properties of zeolites · Application of zeolites

2.1 Zeolites

Zeolites represent a group of more than 50 soft, white aluminosilicate minerals of tectosilicate type, i.e., a three dimensional framework (refer Fig. 2.1a, b) of interconnected tetrahedra, comprising (mostly) of aluminum, silicon and oxygen atoms [1]. They consist of a crystalline structure built from $[AlO_4]^{5-}$ and $[SiO_4]^{4-}$, bonded together in such a way that all four oxygen atoms located at corners of each tetrahedron are shared with adjacent tetrahedral crystals as shown in Fig. 2.1c–e [1 7]. As presented in Fig. 2.1f, if each tetrahedron in the framework contains silicon as its central atom, the overall structure becomes electrically neutral (as in Quartz, SiO_2). In zeolite structures, some of the quadri-charged silicon cations are replaced by triply-charged aluminum, giving rise to a deficiency of positive charge. The so developed charge is balanced by the presence of singly- and/or doubly-charged cations, such as sodium (Na^+), potassium (K^+), calcium (Ca^{2+})and magnesium (Mg^{2+}), elsewhere in the structure, featuring spacious pores or rings [2, 5, 6, 8, 9].

The general formula of a zeolite is $M_{e2/n}\ O \cdot Al_2O_3 \cdot xSiO_2 \cdot yH_2O$ [2, 10], where, M_e is any alkali or alkaline earth atom, n is the charge on that atom, x is the number of Si tetrahedron varying from 2 to 10, and y is the number of water molecules varying from 2 to 7. The Si and Al tetrahedra combinedly form a structural framework in zeolites with centrally located Si or Al atoms and corners

B. Jha and D.N. Singh, *Fly Ash Zeolites*, Advanced Structured Materials 78,
DOI 10.1007/978-981-10-1404-8_2

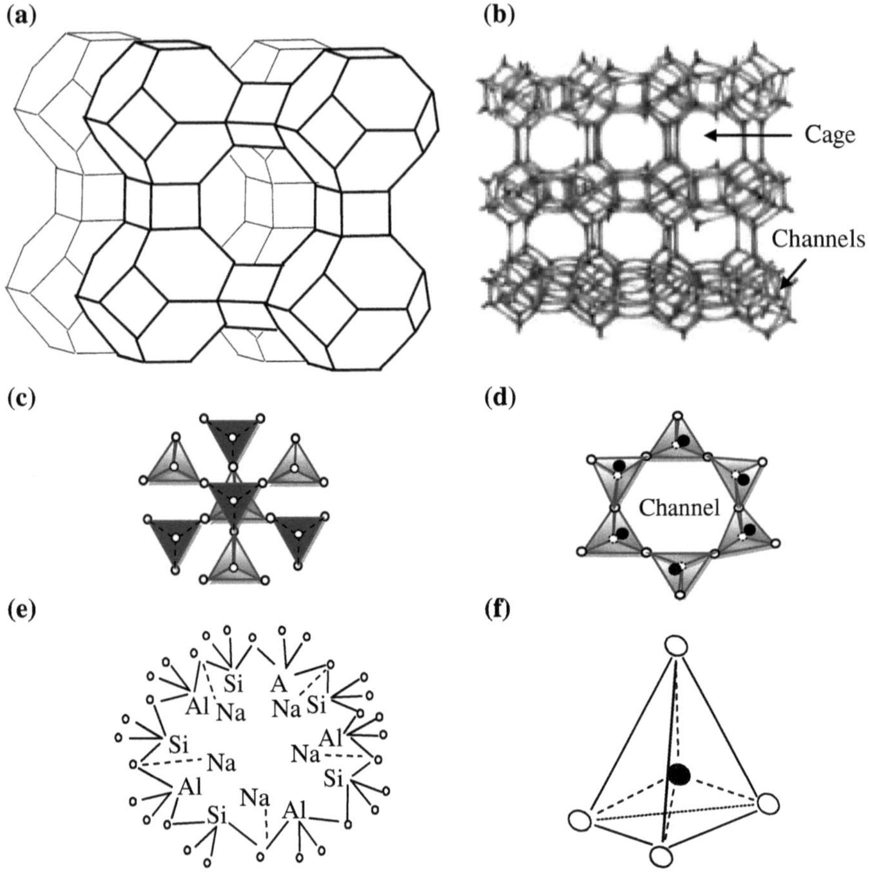

Fig. 2.1 **a** The schematic view of the crystal structure of zeolite 4A and **b** typical zeolite structure showing three dimensional cages and channels [8]. **c** Basic tectosilicate structure of zeolite where *dark* (i.e., vertex in) and *light* (i.e., vertex out) *shades* to add 3-dimensional (3-D) effect and upside down orientation of the tetrahedra for vertex sharing between two rings of the zeolite structure in its 2-D view on a picture plane. **d** Single ring tetrahedron structure and framework of a zeolitic mineral and **e** $[SiO_4]^{4-}$ and $[AlO_4]^{5-}$ in a ring of sodium zeolite and **f** Pictorial representation of a 3-D view of a tetrahedral with centrally located Si or Al atoms, exhibited by *dotted lines* drawn to represent the portion within the body of the mineral [8]

occupied by oxygen atoms. The oxygen atom being common between $[SiO_4]^{4-}$ and $[AlO_4]^{5-}$ tetrahedra, remains oriented in such a way that the framework develops voids or pores in the form of cages and channels between the tetrahedra, as depicted in Fig. 2.1a, b [11, 12]. The structural formula of the zeolite based on its crystal unit cell (assuming both the SiO_2 and AlO_2 as variables) can be represented by $M_{a/n}$ $(AlO_2)_{\mathbf{a}}$ $(SiO_2)_{\mathbf{b}} \cdot wH_2O$, where, w is the number of water molecules per unit cell, and a and b are total numbers of tetrahedra of Al and Si, respectively per unit cell. The ratio of b/a usually varies from 1 to 5, for Mordenite [i.e., Na_8 $(AlO_2)_8$

$(SiO_2)_{40}$, where a is 8, b is 40 and hence b/a is 5] and zeolite 4A [i.e., Na_{96} $(AlO_2)_{96}$ $(SiO_2)_{96}$, where a is 96, b is 96 and b/a is 1]. Exceptionally, some zeolites are having b/a varying from 10 to 100 or even higher than 100 for ZSM −5 type zeolites [6, 8, 10, 13].

As depicted in Fig. 2.1e, a low silica sodium zeolite (Si/Al = 1, Na/Al = 1, i.e., zeolite 4A) possess an open cage within the lattice and a vast network of negatively charged open channels (accommodating Na^+cation) due to presence of the common oxygen atom between Si and Al tetrahedra [12]. In addition, the pores or channels (refer to Fig. 2.1a–c) are of microscopically small size as of molecular dimensions and hence they are also called as the "molecular sieves" which facilitate cation exchange in adsorption process. Based on these attributes, zeolites find applications in separation and filtration processes.

The crystalline lattice structure of zeolites consists of exceptional lattice stability by virtue of which they facilitate considerable freedom of ion-exchange and reversible dehydration. Zeolites can accommodate new cations (mainly sodium, potassium, magnesium and calcium), water molecules and even small organic molecules. Furthermore, ions and molecules in the cages are loosely bound so that they can be removed or exchanged without destroying the zeolitic framework. However, this depends on the chemical composition and the crystalline structures of a specified zeolite. In general, zeolite minerals have been classified into various families as presented in Table 2.1 [2, 7, 8, 14, 15].

Though, zeolites were first identified by Cronsted in 1756 their molecular sieve properties remained untouched until mid 1920s and a lack of development for commercial use of natural zeolites remained for some time more [2, 15]. With this in view, researchers' attention turned to the synthesis of zeolites, popularly known as synthetic zeolites, by employing pure chemicals [15] and/or minerals present in natural resources or their by-products like coal fly ash [3, 4, 16–29].

Table 2.1 Minerals of the zeolite family [8]

Family of zeolites	Minerals	Shape
Analcime	Analcime, pollucite, wairakite, bellbergite, bikitaite, boggsite, brewsterite	Cubic/tetrahedral
Chabazite	Chabazite, willhendersonite, cowlesite, dachiardite, edingtonite, epistilbite, erionite, faujasite, ferrierite	Rhombohedral
Gismondine	Amicite, garronite, gismondine, gobbinsite, gmelinite, gonnardite, goosecreekite	Monoclinic/orthogonal
Harmotome	Harmotome, phillipsite, wellsite	Monoclinic
Heulandite	Clinoptilolite, heulandite, laumontite, levyne, mazzite, merlinoite, montesommaite, mordenite	Monoclinic/orthogonal
Natrolite	Mesolite, natrolite, scolecite, offretite, paranatrolite, paulingite, perlialite	Orthogonal/tetrahedral
Stilbite	Barrerite, stilbite, stellerite, thomsonite, tschernichite, yugawaralite	Monoclinic

2.1.1 Natural Zeolites

Zeolites in nature often, formed as crystals in small cavities of basaltic rocks over the years or as volcanic tuffs or glass altered by the interaction with saline water. These natural zeolites are formed in a number of geological environments such as alkaline deserts, lake sediments, ash ponds and marine sediments at relatively low temperature, under natural conditions. They also get crystallized in geologically young metamorphic rocks in mountainous regions. In the 1950s, geologists discovered that million-ton deposits of volcanic tuff consisting mostly of zeolitic materials are not uncommon [5, 7, 15]. Such zeolitic crystals, by virtue of their unique structures, get filled up with water which can be driven off by heating. As such, the dried up crystals possess a honeycomb-like structure consisting of openings or pores of the order of a few atoms in width (~2–10 Å) [30–32]. The most general formula of natural zeolites is as depicted below [8, 22, 33]:

$$(\mathrm{Li,\ Na,\ K})_p(\mathrm{Mg,\ Ca,\ Sr,\ Ba})_q\left[\mathrm{Al}_{(p+2q)}\mathrm{Si}_{n-(p+2q)}\mathrm{O}_{2n}\right]\cdot \mathrm{m_oH_2O}$$

where, p is the number of monovalent metal ion, q is the number of divalent metal ions, n is the half of the number of oxygen atom and m_o is the number of water molecules.

Natural zeolites such as Clinoptilolite (i.e., popularly known as Clino zeolites) and Chabazite have applications in various diversified fields such as water treatment, fertilizer application for soil amendment and plant growth by establishing better retention of nutrients. The Clinoptilolite has been broadly accepted for its usage in agriculture, soil amendment and feed additives because of its higher acid resistant silica content (viz., Si/Al = 1–5) [14]. However, such zeolites are contaminated by other minerals (e.g., Fe^{2+}, SO_4^{2-}, Quartz, other zeolites, and amorphous glass) and hence they may not be suitable for several important commercial applications where uniformity and purity are essential [8].

2.1.2 Synthetic Zeolites

These zeolites are synthesized by chemical processes, which result in a more uniform and purer state as compared to the natural types in terms of their lattice structures, sizes of pores and cages in their frameworks. The principal raw materials useful for synthesis of synthetic zeolites can be pure chemicals rich in silica and alumina, minerals available on the earth or by-products of industries. Fly ash being an abundantly and cheaply available industrial by-product, rich in minerals containing silica and alumina can be an alternative material for synthesis of synthetic zeolites [20, 34–39]. The type of zeolites formed is a function of the temperature, pressure, concentration of the reagent solutions, pH, process of activation and ageing period, SiO_2 and Al_2O_3 contents of the raw materials [1, 8]. Based on the

Table 2.2 Grades of zeolites [8]

Zeolite grade	Si/Al molar ratio	Some of the common mineral names and their framework codes
Low silica	≤ 2	Analcime (ANA), cancrinite (CAN), Na-X (FAU), natrolite (NAT), phillipsite(PHI), sodalite (SOD)
Intermediate silica	2–5	Chabazite (CHA), faujasite (FAU), mordenite (MOR), Na-Y (FAU)
High silica	>5	ZSM-5(MFI), zeolite-β (BEA)

Si/Al molar ratio in the activated fly ash, zeolites can be classified/graded as "low silica zeolites", "intermediate silica zeolites" and "high silica zeolites", as listed in Table 2.2. In general, for zeolites, an increase in this parameter (i.e., Si/Al from 0.5 to infinity) [5] can significantly result in the increase in various other parameters (viz., acid resistivity, thermal stability and hydrophobicity) except few parameters (viz., hydrophilicity, acid site density and cation concentration) which get decreased [5, 8, 10, 40, 41]. In general, synthetic zeolites hold some key advantages over their counterparts' i.e. natural zeolites. Zeolites type A, X, Y, P and Na-P1 are well known synthetic zeolites synthesized from fly ash which have a wider range of industrial applications than the natural zeolites [1, 8, 20, 22, 36, 42, 43].

2.1.3 Properties of Zeolites

A comprehensive discussion on various properties of zeolites viz., physical, chemical, ion exchange and adsorption properties, mineralogical and morphological characteristics, thermal characteristics, characteristics of zeolites in acidic medium, crystal structure, framework of zeolitic crystals and surface properties is presented in the following.

2.1.4 Physical Properties

The most general physical properties of the zeolites are bulk density and specific gravity (i.e., somewhere in between 2 and 2.4), which can correlate with their porosity (i.e., the measure of the pore volume in zeolite) and the cation exchange capacity (*CEC*) [15, 35, 41, 42].

For example, the observed trends of variations such as one between porosity and *CEC* and another between porosity and specific gravity are exhibited by Fig. 2.2a. It can be noticed that there is negligible change in specific gravity with increase in porosity of zeolites (viz., Analcime, Mordenite, Philipsite, Clinoptilolite, Erionite, Heulandite and Chabazite), whereas, the trend of variation in *CEC* is initially decreasing with increasing porosity up to 34 %. Beyond this, there is reversion in

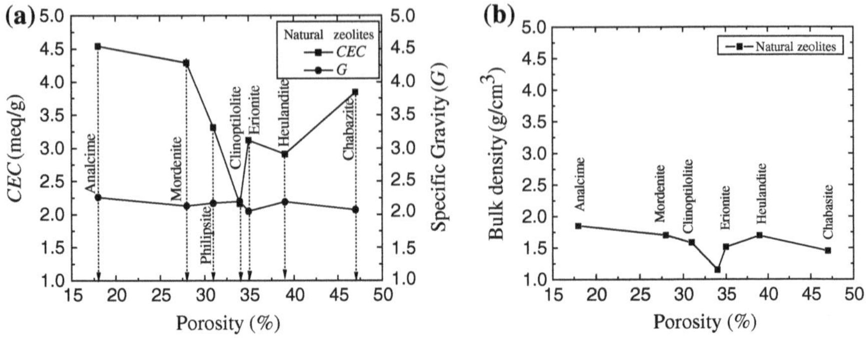

Fig. 2.2 Variation of porosity of natural zeolites **a** with *CEC* and *G*, **b** with bulk density [8]

the trend which can be attributed to the corresponding extreme variations in the purity (i.e., higher grade) of zeolites.

In addition, the surface hardness of zeolite is of the order of 3–5 kg which can be indirectly correlated with its specific gravity [15, 34, 35, 41, 42].

The most common physical property of the ash zeolites is its specific surface area, which is dependent on the extent of dissolution of fly ash particles in alkaline solvents [19, 22, 31]. In line with this, another important physical property of zeolites is their void volume which can directly be correlated with the *CEC* of the synthesized product (e.g., fly ash zeolites, Na-P1) and which in turn depends upon the specific area as depicted in Fig. 2.3a, b [1, 22]. Moreover, both *CEC* and the surface area of the ash zeolites are found to undergo significant variations with increase in molarity and the reaction time, as depicted in Fig. 2.3c, d [22, 29]. From the trends depicted in Fig. 2.3c, it can be observed that the *CEC* increases, marginally, with an increase in concentration, however, the same is noticed to be fluctuating, randomly, with an increase in reaction time. This can be attributed to the variations in the pore size and volume, as depicted in Fig. 2.3a. On the contrary, the surface area maintains an increasing trend with increase in concentration and the reaction time, as depicted in Fig. 2.3d, which can be attributed to increase in dissolution of fly ash ingredients (viz., glass, Quartz and Mullite).

Another important physical property of the ash zeolites is their pore radius R_p. This parameter helps in studying the adsorption properties of zeolites as an adsorbent. R_p can be correlated with the specific surface area SSA_{BET}, which can be determined by nitrogen adsorption technique (i.e., by employing BET method and the relationship, $R_p = 2 \cdot V_p/SSA_{BET}$, where V_p is the pore volume) [44]. The pores are assumed to be cylindrical in shape for natural zeolites; Clinoptilolite and Mordenite, for which SSA_{BET} generally lies between 11–16 m^2 and 115–120 m^2/g, respectively. The trend depicted in Fig. 2.4 exhibits an initial increase in R_p with an increase in SSA_{BET}, up to 20 m^2/g, beyond which it decreases sharply [8]. This trend violates the inverse relationship between the two parameters as mentioned above.

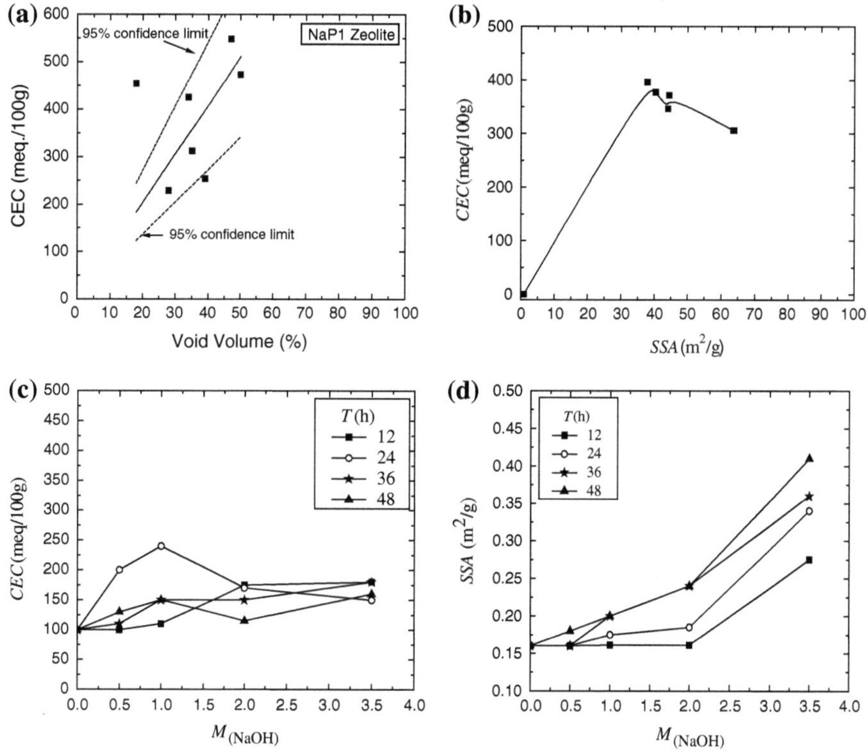

Fig. 2.3 Variation of *CEC* with **a** void volume, **b** specific surface area of the activated fly ash, **c** with concentration of NaOH and **d** variation of specific surface area with concentration of NaOH [8]

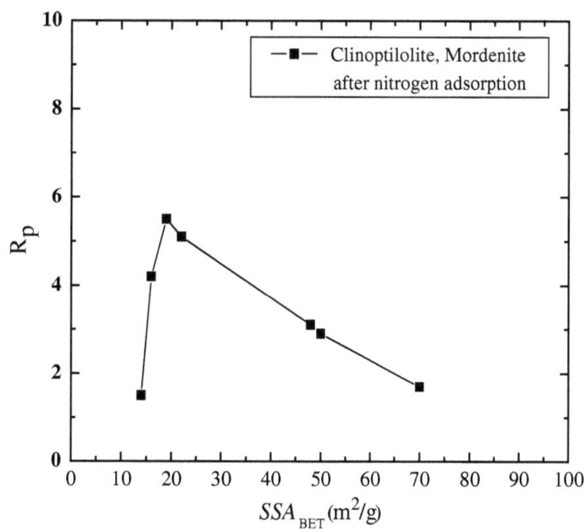

Fig. 2.4 The observed relation between specific surface area and pore radius of zeolite [8]

Table 2.3 Physical properties of the zeolite–clinoptilolite [8]

Bulk density (g/cc)	Clinoptilolite content (%)	CEC (meq./g)	Surface charge density (meq./Å^2)	Si/Al	Pore size diameter (Å)	Pore volume (%)	Total surface area (m^2/g)
2.38–2.81	75–85	0.8–1.2	10×10^{-23}	1–5	4–7	≤ 52	≤ 800

The most general physical property of common zeolites (e.g., Na-X also called zeolite X or Linde X or molecular sieve 13X which is an analogue of natural zeolite popularly known as Faujasite) is their particle size, which has been reported to vary from 2 μm (for bulk-Na-X) to 800 nm (for micro-Na-X) and from 20 to 100 nm for nano-Na-X zeolite [45]. Incidentally, a wide range of variation in the particle size, the effective particle size (i.e., the sieve size which allows passing of 10 % of the material by weight) and the uniformity co-efficient (i.e., the ratio of the sieve size that can permit passage of 60 % of the material by weight to the sieve size corresponding to the passage of 10 % of the material by weight) are quite commonly associated with the zeolites. An example of a commercial grade of the natural zeolite, popularly known as Clinoptilolite, which is used in fertilizers manufactured by St. Cloud, USA, is being cited in Table 2.3 to exhibit a wide range of variations in various physical properties of zeolites [8, 46].

2.1.5 *Chemical Properties*

Zeolites consist of aluminium oxide, calcium oxide, iron oxide, magnesium oxide, potassium oxide, silicon oxide and sodium oxide within their structure with water molecules and/or cations in the pores and the cages [10, 20, 27, 46–48]. A certain fraction of the mass of the zeolites is lost on ignition because of loss of water. Researchers have suggested that, for a material to get zeolited, the ratio of (Si + Al)/O in it should be equal to 0.5 [16, 46–48]. The cation exchange capacity (*CEC*), adsorption properties, pH, and loss on acid immersion of zeolites are some of the chemical properties which are reported to depend on the chemical composition of the synthesized products. Table 2.4 presents typical chemical composition of a fly ash, its crystalline constituents (viz., Quartz and Mullite), one commercial grade synthetic zeolite, a fly ash zeolite and their comparison with a natural zeolite [47, 48].

It can be noticed from the data presented in Table 2.4 that the chemical composition of the fly ash zeolites (i.e., synthesized by Ojha et al. [48] and Park et al. [47]) is very close to the commercial grade synthetic zeolite 13X with Si/Al ratio equal to 1.5 [12, 47, 48], whereas, natural zeolite is comparatively rich in silica with Si/Al ratio equal to 4 [12, 21]. Hence, it can be opined that a wide range of chemical

Table 2.4 Chemical composition of fly ash, its minerals and zeolites of natural and synthetic types (by weight%) [8]

Material	Oxide (%)						
	SiO_2	Al_2O_3	Fe_2O_3	TiO_2	CaO	Na_2O	K_2O
Fly ash	52.1	32.1	5.5	2.1	0.75	1.9	1.3
Mullite	27.8	71.5	–	–	–	–	–
Quartz	>99	–	–	–	–	–	–
Fly ash zeolite	43.6	29.5	3.6	1.9	0.7	20.5	0.91
Zeolite-13X (commercial grade)	48.26	31.85	3.2	0.08	0.38	15.7	0.07
Natural zeolites	64.0	16.1	2.8	0.3	0.2	3.5	3.7

– not applicable

transformation takes place from mineral phase of the fly ash to the corresponding fly ash zeolite phase.

2.1.6 Ion Exchange and Adsorption Properties

Zeolites usually gain cations (viz., Na^+, K^+ and NH_4^+)during the synthesis process or by interaction with the surrounding medium by virtue of their ion exchange or adsorption characteristics [1, 11, 17, 29, 31, 39, 49]. In fact, the cations are accommodated to balance the negative charge developed on the surfaces of pores in zeolites. This can be attributed to the replacement of Si atom by Al atom in some of the $[SiO_4]^{4-}$ tetrahedra and its conversion into the $[AlO_4]^{5-}$tetrahedron which is interconnected to other $[SiO_4]^{4-}$ tetrahedron by common oxygen atom as depicted above in Fig. 2.1e, f. As for example, ion exchange process can be described by exposing a sodium zeolite to a waste water sample or a fresh solution containing other metal cations (e.g. NH_4^+). In fact, the sodium ions of the zeolite can be exchanged by ammoniumions provided they are not excluded from the zeolite pores due to higher molecular size. Based on the findings of the previous researchers, a typical ion exchange process of waste water treatment by zeolite application at room temperature is simulated below by allowing ammonium chloride solution to pass through a zeolite sample, as depicted in Fig. 2.5. It can be noticed that the Na^+ of the zeolite can easily be exchanged with the NH_4^+ by this process of ion exchange.

The heavy metal cations such as Rb, Cs, Ag, Cd, Pb, Zn, Ba, Sr, Cu, Hg, Co, and Cr have affinity towards zeolites, although, their selectivity by the zeolites for exchange depends on the hydrated molecular size of the cations, their relative concentrations in the medium associated with the process and the Si/Al molar ratio of the zeolite framework [3, 25, 37, 38, 43, 50, 51].

Based on these properties, the zeolites have been also found to adsorb gases and separate them for useful industrial applications. The most common gases being CO,

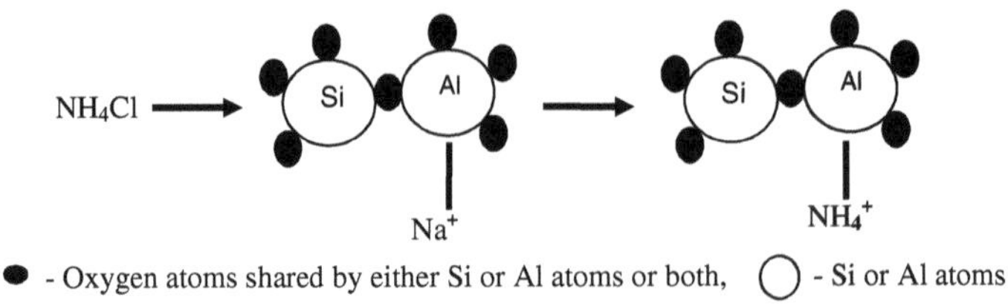

Fig. 2.5 Typical ion exchange process in a mixture of ammonium chloride and any sodium zeolites [8]

CO_2, SO_2, H_2S, NH_3, HCHO, Ar, O_2, N_2, H_2O, He, H_2, Kr, Xe, CH_2OH, Freon and Formaldehyde [38, 43, 51, 52].

2.1.7 Mineralogical Properties

X-ray diffraction (XRD) analysis has been a useful tool to check the presence of minerals (viz., Mullite, Hematite, Magnetite and α-Quartz) as the main crystalline phase in the fly ash and its zeolites, in addition to the presence of amorphous glassy phase [16, 38]. Furthermore, micrographs obtained by scanning electron microscopy (SEM) of the fly ash and its zeolites, as depicted in Fig. 2.6a, have been found to be a useful tool for demonstrating the shape and grain size of constituent minerals (refer Table 2.5 [8, 24]).

However, after crystallization, only the new peaks detected in the XRD pattern can depict the presence of zeolite crystals, of varying intensity corresponding to different reaction times for treatment with NaOH, as depicted in Table 2.6 [4, 8, 20, 37]. It can be noticed that the XRD intensities reduce significantly, corresponding to an increase in the reaction time from 2.5 to 3.0 h. This can be attributed to an increase in dissolution of crystals of zeolite P and Quartz, which is an indication of their less stable forms as compared to Mullite. Further, based on the location of peaks, the maximum intensities for some common zeolites are presented in Table 2.7. From the data presented in the table, it can be noted that most commonly occurring ash zeolites (viz., zeolite P, Na-A and Na-X) can easily be identified by the appearance of new peaks in the XRD diffractogram.

Similarly, XRD of naturally available zeolitic minerals do exhibit variations in their mineralogical features, and hence their properties (viz., structure type, pore size dimensions, channel dimensions, surface charge density and electro- negativity), which are the decisive factors related to zeolites and their industrial applications [16, 23, 53, 54].

Fig. 2.6 SEM micrographs of the **a** Fly ash (spherical morphology), **b** Na-A (cubic morphology), **c** Sodalite (ball shaped morphology), **d** Cancrinite (Hexagonal, prismatic, needle like morphology), **e** Zeolite Y (cubic morphology), **f** Zeolite Na-X (octahedral morphology) [8]. **g** Mordenite (acicular or prismatic crystals) and Analcime (spherical crystals), **h** Clinoptilolite (hexagonal) and NaP1 (ball shaped), **i** Fibrous Na-P1, **j** Na-P1 (polycrystalline), **k** Na-A with emerging agglomerates and **l** Na-X (Cubic) [8]

2.1.8 Morphological Properties

The size and shape of crystals of minerals can be ascertained by interpreting the SEM micrographs of the raw materials (viz., fly ash) and the end products obtained from the zeolitization process [8]. The SEM micrographs of fly ash reveal the presence of spherical particles of size 50–80 µm along with broken hollow spheres,

Fig. 2.6 (continued)

as depicted in Fig. 2.6a. However, after zeolitization, most of the products retain the initial spherical morphology of the fly ash, except some surface alterations leading to roughness due to chemical action as shown in Fig. 2.6b, c. In addition, needle-like crystals of different sizes can be observed on the grain surface, which can be attributed to the growth of Cancrinite crystals (refer Fig. 2.6d). Such characteristic hollow structures of zeolitic minerals facilitate their application in several industrial processes and products [8, 54].

Three dimensional surface pores of small crystals of zeolites A, X and Y can be observed in the SEM images, as depicted by black spots in Fig. 2.6b, e, f [3, 8], whereas; one dimensional large crystals can be seen in the zeolites, Mordenite, as

Table 2.5 Particle shapes and size of minerals present in fly ash and its zeolites [8]

Minerals		Particle shapes	Size (μm)
Source	Type		
Fly ash	Hematite	Spheroidal aggregates	5–10
	Magnetite	Spherical form	5–15
	Glass	Spherical form, spheroidal mode, yellowish and greenish	10–400
	Quartz	Sharp edged, elongated, oval and spherical	20–250
	Carbon	Irregular and jagged form	22–150
	Mullite	Pillar like, single form with cleavage and colorless	60–420
Zeolites	Na-P1	Fibrous	–
	Analcime	Tetrahedral	10–20
	Na-X	Octahedral	20
	Cancrinite sodalite	Hexagonal Hexagonal	20–30
	Na-A	Cubic	40–50

– not applicable

Table 2.6 Effect of reaction time on the XRD intensity of the minerals [8]

Reaction time (h)	XRD peak intensity (counts per second)		
	Zeolite P	Quartz	Mullite
0	325	260	85
0.5	425	215	82
1.0	515	240	78
1.5	500	190	75
2.0	550	165	70
2.5	570	160	65
3.0	550	140	60

Table 2.7 Maximum peak intensities and their location in the X-ray diffract gram for common ash zeolites [8]

Zeolite	2θ (°, Cu Kα)	Peak XRD intensity (counts per second)
P	28	650
Na-A	7	650
X	6	1800

depicted in Fig. 2.6g [8, 34, 55]. It has been reported that the initial ball-shaped morphology as seen in Fig. 2.6c, h of zeolite Na-P1 get transformed to star-shaped grains or other crystal shapes after continuous dissolution and crystallizationas depicted in Figs. 2.6h, j, k, l, p [8, 30].

Natural zeolites which consist of fibrous morphology include Natrolite, Tetranatrolite, Paranatrolite, Mesolite, Scolecite, Thomsonite, Erionite and

Mordenite are also depicted in Fig. 2.6i. Clinoptilolite has been reported to occur as idiomorphic plates and laths, which are several micrometers in length and 1–2 μm in thickness, whereas, most of the crystals display characteristic monoclinic symmetry and many are coffin-shaped. Some zeolites appear as fibrous aggregates as depicted in Fig. 2.6i, whereas, others as robust, non-fibrous crystals as seen in Fig. 2.6b, d, h, m, n [8, 34, 42, 54, 56, 57].

2.1.9 Thermal Characteristics of Zeolites

The thermal properties (viz., temperature resistance, thermal stability, thermal conductivity and heat capacity of zeolites have been studied to investigate the loss of water or thermally induced cracking at higher temperatures [46]. The thermal stability of zeolites has been noticed to increase with their crystallinity. A higher SiO_2/Al_2O_3 ratio and *CEC* of the zeolites can directly add to their temperature resistance. The synthesized Na-X zeolite has been found to lose its crystallinity between 973 and 1073 K [27]. High-silica zeolites (HSZ) are thermally and chemically stable up to about 1000 °C whereas; pure-silica zeolites are stable in all mineral acids except hydrofluoric acid. HSZs have been found to have Si/Al ratios greater than 50, while the ratio for low-silica zeolite (LSZ) is less than 2–5. The variation of the degree of thermal stability of fly ash zeolites, has been found to follow the trend Na-P1, < Na-X < Sodalite. Moreover, Na-P1, Na-X and Sodalite, have been opined to maintain their crystalline structure for temperatures below 300, 700 and 900 °C, respectively [58].

The thermal conductivity and heat capacity of zeolite 4A ($Na_{96}Al_{96}Si_{96}O_{384}$, popularly identified as Linde A, or Na zeolite) have been reported by earlier researchers [58], for the temperature range 35–300 K, who have demonstrated that the thermodynamic stability of zeolites depends on the strength of the Si-O and Al-O bonds in their structural framework [11]. It has also been observed that such stability of zeolites gets enhanced by an increase in its Al content. Further thermal conductivity of the zeolite Na-X, can be affected by its particle size (2 μm to 800 nm), the temperature range (5–390 K), and the degree of packing and the distribution of voids. The inter relationship between thermal conductivity (k) of zeolite Na-X and the temperature is presented in Fig. 2.7 [8, 45].

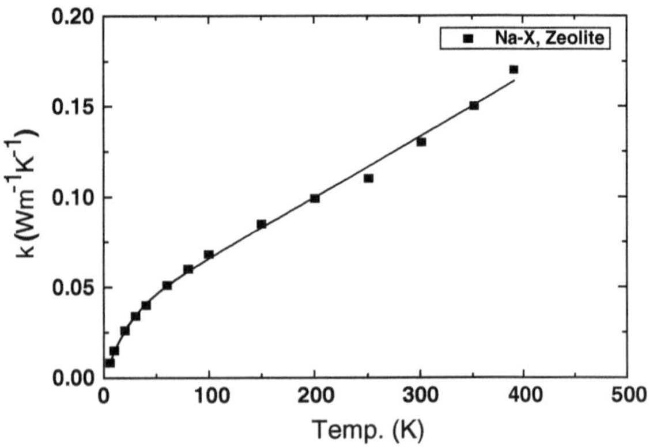

Fig. 2.7 Effect of temperature on thermal conductivity of Na-X zeolite [8]

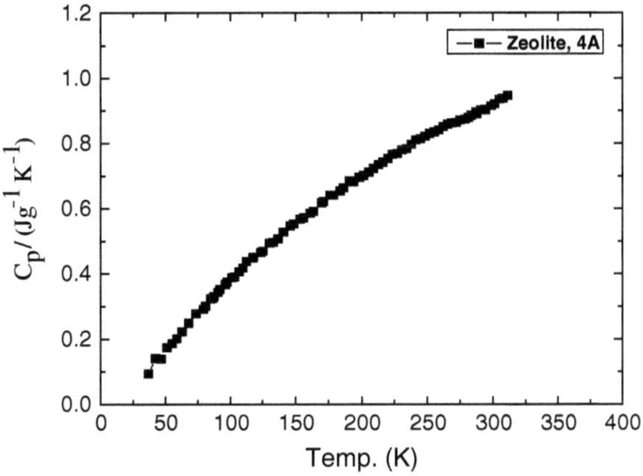

Fig. 2.8 Effect of temperature on heat capacity of zeolite 4A [8]

Incidentally, the heat capacity of zeolites has been reported to be a useful tool to demonstrate the interdependency of their thermodynamic stability and phase transition with their structure, under varying temperature, as depicted in Fig. 2.8 [8, 11].

2.1.10 Stability of Zeolites in Acidic Medium

Zeolites are composed of various atoms of different electronegativity (refer to Table 2.8). However, their intermediate electronegativity (i.e., geometric mean of the component atoms after redistribution of the electrons in the compound) is reported to be established as demonstrated by Sanderson's principle of electronegativity equalization (i.e., when two or more atoms initially different in electronegativity combine chemically, they become adjusted to the same electronegativity within the compound). The intermediate electronegativity can be determined for a compound, P_u, Q_v, R_z by Eq. (2.1) [59]:

Table 2.8 Electronegativity (*E*) of elements [8, 59]

Element	*E*	Element	*E*
H	3.55	K	0.42
C	3.79	Ca	1.22
N	4.49	Zn	2.98
O	5.21	Rb	0.36
F	5.75	Sr	1.06
Na	0.70	Cd	2.59
Mg	1.56	Cs	0.28
Al	2.22	Ba	0.78
Si	2.84	Ag	2.57
Ge	3.59	–	–

– not detected

$$E_i = \left[(E_P)^u.(E_Q)^v.(E_R)^z\right]^{1/(u+v+z)} \tag{2.1}$$

The change in electronegativity undergone by each atom can be obtained by Eq. (2.2).

$$\Delta E_P = 2.08.\sqrt{E_P} \tag{2.2}$$

The partial charge on the atom P can be obtained by Eq. (2.3).

$$[E_i - E_P]/\Delta E_P \tag{2.3}$$

where, E is the electronegativity of the atom (refer Table 2.8); P, Q and R are the atoms in the mineral, and, u, v, and z are their numbers in one mole of their compound, respectively.

With this in view, the acidic strength of zeolites with reference to the stability of their framework and the presence of H^+ proton sites in the crystal structure of zeolites can be directly correlated with their intermediate electronegativity.

With the intention of establishing the effect of the variation in bonding and structure of zeolitic mineral on its acidic strength, a model applicable for the bond angles between various elements is being depicted in Fig. 2.9 [8, 60]. In fact, the angle β (refer Fig. 2.9) between Si-O-Al bonds in the zeolite crystal structure can play an important role against its surface corrosion in acidic medium. The high T-O-T (i.e., T stands for Si and Al atoms, and O stands for oxygen atom) bond strength has been noticed in high silica zeolites, whereas, it has been demonstrated to be low in case of low silica zeolite, e.g., Faujasite. In addition, the O-H bond might become unstable due to infra-red radiations, at lower bending frequency, in a zeolite structure. The probability of instability of the bond -[Si-O-Al]- has been found to be more corresponding to higher value of angle, β ($\approx$127°) between the bond, as depicted in Fig. 2.9 [8, 59].

Similarly, the Si-O and Al-O bond lengths are also reported to be dependent on -[Si-O-Al]- bond angle. With this in view, it can be opined that an increase in β results in decrease in θ and hence lowering of the acidic strength. It has been demonstrated [8, 59] that lower the required frequency for stretching the OH bond, the greater would be the acidic strength of zeolites. For an example, it has been

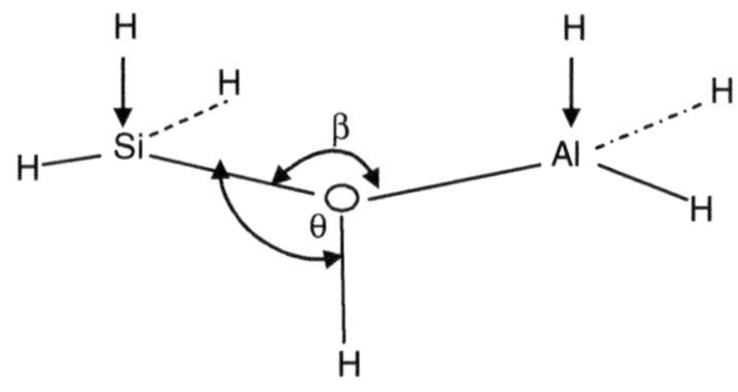

Fig. 2.9 A model of bond angle Si-O-Al and Si-O-H in the zeolite framework [8, 59]

reported that the bending frequency of Mordenite is higher than that of Faujasite zeolites [8, 59].

2.1.11 Crystal Structure of the Zeolite

Zeolites, as minerals of the tectosilicate group, i.e., three dimensional arrays of interconnected SiO_4 tetrahedra, have basically three different structural variations [8, 15]:

(a) Chain like structures: the crystals appear as acicular or needle like prismatic crystals as of Natrolite ($Na_2Al_2Si_3O_{10} \cdot 2H_2O$). Such zeolite (specific gravity, $G = 2.2$; hardness $H = 5.5$) can appear as compact fibrous aggregates with fibers of divergent radial arrangement as depicted in Fig. 2.10a, b. In Fig. 2.10b, $[SiO_4]^{4-}$ and $[AlO_4]^{5-}$ tetrahedra have been depicted as shaded and non-shaded triangles (white in color), respectively, while the shaded small circles depict the vertex of the triangles, i.e., the oxygen atom. Moreover, small sized circles between the vertexes denote Na^+, as external linkage to the tetrahedral and exhibited in Fig. 2.10b [8].

Fig. 2.10 Crystal structures of common zeolites **a**, **b** Natrolite, **c** Heulandite and **d** Chabazite [8]

(b) Sheet like structures: the crystals appear as flattened, platy or tubular usually with good basal cleavages of Heulandites [(Ca, Na)$_{2-3}$, Al$_3$(Al, Si)$_2$ Si$_{13}$ O$_{36}$ · 12H$_2$O] with ends like wedges. The crust of such crystals (G = 2.2, H = 3–4 kg) has appearance like rhombic prisms as depicted in Fig. 2.10c [8].
(c) Framework structures: the crystals appear more equal in dimensions as that for Chabazite [(Ca Na$_2$ K$_2$ Mg) Al$_2$ Si$_4$ O$_{12}$ · 6H$_2$O], which has rhombohedral/cubic shaped crystals (G = 2.0 to 2.2, H = 3–5 kg), as depicted in Fig. 2.10d [8].

Keeping the above structures in view, the natural zeolites have a unique three-dimensional honeycomb structure (Fig. 2.6i), which creates an open and negatively charged framework through which liquid and gases can be exchanged or absorbed.

2.1.12 Framework Structure of Zeolitic Crystals

Each of the silicon and aluminium tetrahedra (refer Fig. 2.11), present in a zeolite, is popularly known as its primary building unit (PBU). Whereas, the simple geometrical shapes (i.e., ring shapes, designated by R in Fig. 2.11) are created by inter linkage between two or more tetrahedra. Hence, the formed linkages are called secondary building units (SBU). To clarify this fact, such units depicted in Fig. 2.11a–g) have been picturized as single and double rings, respectively. It can be noticed from these figures that each ring is composed of four-, five-, six- or eight-numbers of inter-linked several Si and/or Al tetrahedra, represented by small circles (refer Fig. 2.1c, d, f, where these tetrahedra have been represented by actual tetrahedral shapes) [5, 8].

It can be noticed that, each SBU consists of the lines representing oxygen bridges (-O-), whereas intersection points of these lines represent the positions of silicon or aluminium atoms. As a result, -[Si-O-Al]- linkages are formed which can develop a specific geometry of the surface pores in the zeolite framework. Further, regular cavities of discrete shape and size can get enclosed in between the linkages of various SBUs, which can accommodate different cations (viz., Na, K, Ca, Li and Mg) as an ion exchange or a molecular sieve [5].

For example, the shape and dimension of channels and/or pores of some common zeolites are being depicted in Fig. 2.11h, i, j [8]. It has been established that Si/Al ratios have considerable effect on other properties (viz., *CEC*, channel dimension, void volume and specific gravity). Figure 2.11h represents the zeolite, Analcime (ANA)-distorted 8-ring, viewed along [110], which has cubic shaped irregular channels of size (Å): 4.2 × 1.6 [8].

Figure 2.11i depicts Heulandite (Clinoptilolite group: HEU), 8 membered ring, along [001], which has monoclinic crystals and 2-dimensional channels of size (Å) 4.6 × 3.6, 4.7 × 2.8 and 7.5 × 3.1, which is variable due to considerable

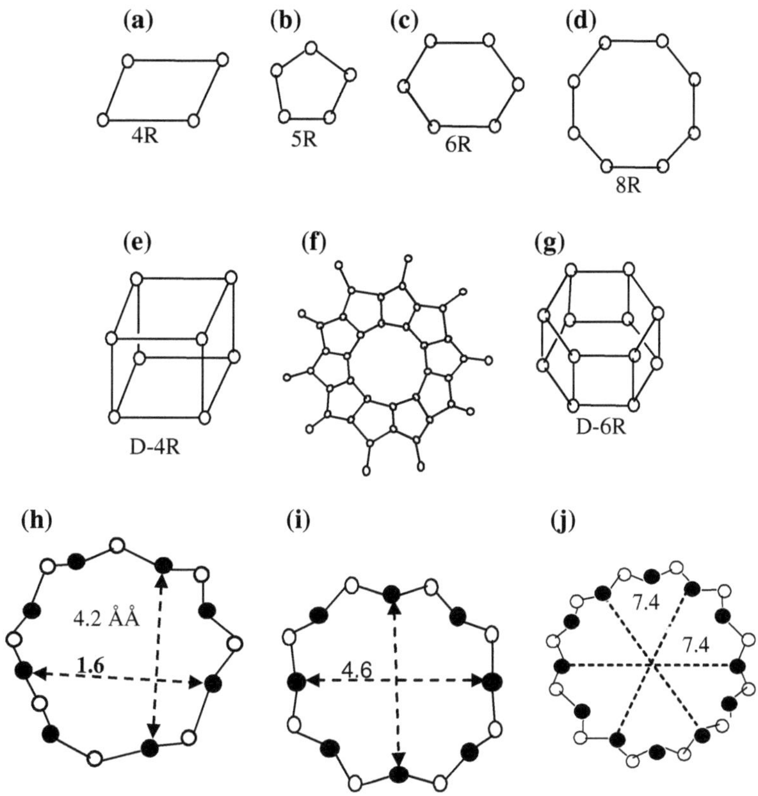

Fig. 2.11 Different types of linkages of tetrahedra in the secondary building units of framework structures of zeolite groups. **a**, **c**, **d** Analcime group, **b** Heulandite and Mordenite groups, **e** Phillipsite group, **f** Pentasil and **g** Chabazite group [8]. Variation in channel shapes and dimensions of common zeolites. **h** Analcime—8R, **i** Clinoptilolite—8R and **j** Faujasite—12R [8]

flexibility of the framework. Figure 2.11j depicts the zeolite Faujasite: FAU-12 ring, viewed along [111], which has 3-dimensional channels of size 7.4 Å [8, 12, 47].

Based on such variations in the framework of zeolites, a 'structure code' has been assigned to each one of them, for the sake of simplicity in their identification. In fact, as many as 191 types of structure codes (i.e., Framework Type Code, FTC) have been proposed by the Structure Commission of the International Zeolite Association (IZA-SC) [5]. It is notable that several zeolites exhibit similarities in their structures which can be grouped together to form a iso-structural group of zeolites. Table 2.9 represents details of the common groups based on the type of structure of zeolites [8].

Further, the frame work structure of zeolite can be correlated with its Si/Al ratio. It has been observed that, in general, with an increase in the Si/Al ratio, the zeolite structure gets transformed from 4-, 6- and 8-membered rings to 5-membered rings

Table 2.9 Different types of framework structure and their iso-structural species [8]

Zeolite	Structure	Main species	FTC	Isostructural species
Analcime	4R and 6R	Analcime	ANA	Ca-D, Kehoeite, Leucite, Na-B, Pollucite, Viscite, Wairakite
Natrolite	Chain of PBU	Natrolite Edingtonite Thomsonite	NAT EDI THO	Laubanite, Mesolite, Metanatrolite, Scolecite K-F Gonnardite
Chabazite	D-6R	Chabazite Cancrinite Erionite Gmelinite Levyne Losod Linde L Mazzite Offretite Offretite Sodalite	CHA CAN ERI GME LEV LOS LTL MAX MAX OFF SOD	Linde D, Herschelite, Linde R Basic Cancrinite – Linde S, Na-S Levynite, ZK-20, LZ-132, NU-3 – – Omega, ZSM-4 Zeolite O Basic Sodalite, Danalite, Nosean Hydroxysodalite, Sodalite hydrate,
Phillipsite	D-4R	Phillipsite Li-A Gismondine Merlinoite	PHI ABW GIS MER	Harmotone, Wellsite, ZK-19 $CsAlSiO_4$, $RbAlSiO_4$ Amicte, Garronite, Linde B, Na-P1, P, P_c, P_t, Na-P K-M, Linde W
Heulandite	4R or 5R	Heulandite Brewsterite Stilbite	HEU BRE STI	Clinoptilolite – –
Mordenite	5R	Mordenite Ferrierite Bikitaite Dachiardite Epistilbite	MOR FER BIK DAC EPI	Na-D, Ptilolite, Zeolon Sr-D, ZSM-21, ZSM-35, ZSM-38 – – –
Faujasite	Cubic	Faujasite ZK-5 Linde A	FAU KFI LTA	Linde X, Y, ZSM-20, Ba-P, Ba-Q, P-[Cl], Q-[Br], ZK-4, ZK-21, ZK-22, Alpha, N-A

LTA Na-A or Linde A zeolite (with Si/Al = 1 and SBU ≈ D-4R)
FAU Na-X and Na-Y zeolites (with Si/Al = 1.23 and 2.5, respectively and SBU ≈ D-4R)
CHA zeolite Chabazite (with Si/Al = 2 and SBU ≈ D-6R)
HEU zeolite Clinoptilolite (with Si/Al = 5 and SBU ≈ 5R)
ERI zeolite Erionite (with Si/Al = 3 and SBU ≈ D-6R)
ANA zeolite Analcime (with Si/Al = 2, SBU ≈ 4R-6R)
MOR zeolite Mordenite (with Si/Al = 5 and SBU ≈ 5R)
– not available

[5]. For example, a series of ash zeolites with their structure code, corresponding to their Si/Al ratio, and SBU (shown in paranthesis) have been listed, in the footnote of Table 2.9.

2.1.13 Surface Properties

The surface properties (viz., hydrophobicity, hydrophilicity and binding to reactant molecules) of zeolites bearing negative surface charge can be varied by organic functionalization of their internal and external surfaces, which can improve their affinity to absorb water and other cations. The zeolites of a particular pore size on their external surface can allow penetration of molecules of smaller size or shape to their internal pores by diffusion. The organic cations have been found too large in size to enter the internal pores and hence they are adsorbed in the surface pores of zeolites. Furthermore, Si/Al ratio is an important parameter, which can influence such adsorption by zeolites. This is based on the fact that more the number of aluminium atoms, more will be the electronegativity of the zeolite pore surfaces which correspond to less Si/Al ratio. For example, zeolites Na-A possesses lower Si/Al (~ 1) molar ratio as compared to zeolites X and Y. As such, it can compensate the exchange of Si by Al atoms in the framework to avoid formation of Al-O-Al linkage, with extra framework cations on its internal and external surfaces both, to maintain its uniform acidic strength and intermediate elecronegativity [8, 59].

In order to modify the surface features, the zeolites can be treated with long chain type surfactants, such as hexa-decyl-tri-methyl-ammonium chloride (HDTMA), stearyl-dimethyl-benzyl-ammonium chloride (SDMBA) and distearyl-dimethyl-ammonium chloride (DSDMA). As a result, they have been found to replace inorganic cations like Na^+ and Ca^{2+} from the external surface of zeolites [8, 61]. Moreover, surface modification of synthetic zeolites: A, X, Y and the natural zeolite, Clinoptilolite has been reported to occur by action of cationic surfactants (viz., HDTMA chloride, SDMBA chloride and DSDMA chloride) on the surface of the zeolites [8, 43, 62].

Such modification results in an alteration in the surface properties so much that the hydrophilic zeolites (i.e., with Si/Al < 10) are converted into hydrophobic zeolites (i.e., with Si/Al > 10) which can absorb molecular diameters (e.g., organic cations) larger then water [5, 8]. With this in view, it has been reported that the adsorption capacities of different surface modified zeolites can increase with the increase in their Si/Al ratio [8]. This can be attributed to the corresponding increase in their uniform pore size and their adsorption capacities which can follow the increasing order such as CLI > Ca-Y > Ca-X > Ca-A > Na-Y > Na-X > Na-A as presented in Table 2.10 [8, 62]. The adsorption capacity of zeolites can become a tool for proving the superiority of one zeolite over another for application point of view e.g., removal of pesticides from the environment [8, 62].

Table 2.10 Maximum adsorption capacity of common zeolites for organic cations [8]

Zeolite	Si/Al	Pore size (Å)	Adsorption capacity of the organic cations (μmol/g)			
			MB	HDTMA	SDMBA	DSDMA
Na-A	1.00	4.2	14	61	50	29
Ca-A	1.00	4.9	27	160	91	61
Na-X	1.23	7.4	24	84	70	55
Ca-X	1.23	7.6	63	192	101	72
Na-Y	2.50	7.9	52	116	70	60
Ca-Y	2.50	8.0	86	208	115	77
CLI	5.00	4.4	148	388	288	158

2.1.14 Critical Evaluation of Properties of Some Commonly Available Zeolites

Zeolites are identified by their most critical and valuable property, known as the cation exchange capacity (*CEC*), which defines its suitability for various industrial applications. Apart from this, various attributing characteristics (viz., pore diameter and pore volume) of pores or channels, specific gravity and particle compositions (viz., chemical and mineralogical), particle shapes and size (i.e., morphology) of the zeolites need to be critically evaluated for fixing their suitability for a specific application. The pores in the zeolite add to their values as molecular sieve for separation of particular type of fluids and gases. Moreover, the particle shapes and sizes, mineralogy and morphology of the zeolites can vary a lot depending up on the complexities involved in the chemistry of their synthesis. As such, the particle sizes can be grouped separately as meso-porous sizes (10–60 μm) and micro-porous sizes (<10 μm) [24], which can have direct and/or indirect correlations with *CEC*, specific gravity and other properties of zeolites. In general, the specific gravity of zeolites has been reported to vary from 2 to 3 [20]. In order to present a simplified picture of some zeolites, their comparison with respect to others, and to fix the superiority of a particular zeolite over other, Table 2.11 presents various properties of fly ash zeolites which can be useful for the readers [2, 8, 20, 24, 25, 29, 51, 52, 59].

Table 2.11 Properties of some commonly available fly ash zeolites [8]

Zeolite	*CEC* (meq./g)	Channel/pore diameter (Å)	Pore volume (%)	Specific gravity	Chemical formula
X/Linde X	5	7.3	50	1.93	$NaAlSi_{1.23}O_{4.46} \cdot 3.07\ H_2O$
Clinoptilolite-8R	2.54	3.9–5.4	39	2.16	$(Na, K)\ Al_3\ Si_5O_{12} \cdot 3H_2O$
Laumontite	4.25	4.6–6.	34	2.20–2.30	$CaAl_2Si_4O_{12} \cdot 4H_2O$
Erionite	3.12	3.6–5.2	35	2.02–2.08	$(Na_2,Ca_6,\ K)_9Al_9Si_{27}O_{27} \cdot 27H_2O$
Mordenite 8R	2.29	2.6–5.7	28	2.12–2.15	$Na_8\ (Al_8Si_{40}O_{96}) \cdot 24H_2O$
Hydroxysodalite	0.3	2.3	–	–	$Na_4Al_3Si_3O_{12} \cdot OH$
A/Linde A	4.7	4.1	47	1.99	$Na_{12}Al_{12}Si_{12}O_{48} \cdot 27\ H_2O$
Y	3.12	7.3	–	–	$Na_{1.88}Al_2Si_{4.8}O_{13.54} \cdot 9H_2O$
Na-P1	2.7	3.1 × 4.5/ 2.8 × 4.8	–	–	$Na_6Al_6\ Si_{10}O_{32} \cdot 12H_2O$
Analcime	0.6–4.54	2.3–2.6	18	2.24–2.29	$NaAlSi_2O_6 \cdot H_2O$
Chabazite	3.81	3.7–4.2	47	2.05–2.10	$CaAl_2Si_4O_{12} \cdot 6H_2O$
Herschelite	2.1	3.5	NA	NA	$Na_{1.08}Al_2Si_{1.68}O_{7.44} \cdot 1.8H_2O$
KM- zeolite	1.9	4.5	NA	NA	$K_2Al_2Si_3O_{10} \cdot H_2O$
Faujasite	1.41	7.4		1.98	$Na_2Al_2Si_{3.3}O_{8.8} \cdot 6.7H_2O$
Hydroxycancrinite	–	–	–	–	$Na_{14}Al_{12}Si_{13}O_{51} \cdot 6H_2O$
Tobermorite	–	–	–	–	$Ca_5(OH)_2Si_6O_{16} \cdot 4H_2O$
Linde F	1.9	–	–	–	$KAl\ SiO_4 \cdot 1.5H2O$
Parlialite	–	–	–	–	$K_9NaCaAl_{12}Si_{24}O_{72} \cdot 15H_2O$
Kasilite	–	–	–	–	$KAlSiO_4$
Phillipsite-8R	–	4.2–4.4	–	2.2	$(CaNa_2K_2)_3 \cdot Al_6Si_{10}O_{32} \cdot 12H_2O$
Sodalite	0.3	2.3	–	2.27–2.33	$Na_4Al_3(SiO_4)_3Cl$

– not available

2.2 Conclusions

Zeolites are tetrahedral crystalline aluminosilicate minerals with a general chemical formula (i.e., in terms of oxides of Si and Al), represented as $M_{e2/n}$ O · $Al_2O_3 \cdot xSiO_2 \cdot yH_2O$. Moreover, a generalized structural formula (i.e., in terms of AlO_2 and SiO_2 both assumed as variable parameter) of a crystal unit cell of zeolites correspond to $M_{a/n}$ $[(AlO_2)_{\mathbf{a}} (SiO_2)_{\mathbf{b}}] \cdot wH_2O$. The ratio of b/a usually varies from 1 to 5, for common zeolites, Mordenite and zeolite 4A, respectively. Zeolites have also been synthesized when b/a varies from 10 to 100 or even higher, as in case of ZSM-5 type of zeolites. Most important properties of zeolites are their *CEC*, channel or pore diameters and volume, specific gravity, chemical composition, mineralogical and morphological characteristics. The presence of negatively charged microscopically small sized surface and/or internal pores of zeolites adds to their values as absorbent and/or adsorbent. They are also called as the "molecular sieves" as cation exchange materials. Based on these attributes, zeolites find applications in separation and filtration processes.

References

1. Scott, J., Guang, D., Naeramitmarnsuk, K., Thabuot, M., Amal, R.: Zeolite synthesis from coal fly ash for the removal of lead ions from aqueous solution. J. Chem. Technol. Biotechnol. **77**, 63–69 (2001)
2. Zeolites other than erionite. IARC Monograph. http://monographs.iarc.fr/ENG/Monographs/vol68/mono68-11.pdf (1997). Accessed 04 Sept 2015
3. Rayalu, S., Meshram, S.U., Hasan, M.Z.: Highly crystalline faujasitic zeolites from fly ash. J. Hazard. Mater. **B77**, 123–131 (2000)
4. Murayama, N., Yamamoto, H., Shibata, J.: Mechanism of zeolite synthesis from coal fly ash by alkali hydrothermal reaction. Inter. J. Miner. Process. **64**, 1–17 (2002)
5. Scott, M.A., Kathleen, A.C., Dutta, P.K.: Handbook of zeolite science and technology. CRC Press, New York (2003). ISBN 0-8247-4020-3
6. Elliot, A.D., Zhang, D.K.: Controlled release fertilizers, a value added product produced from fly ash. Cooperative research Centre for coal in sustainable development, Centre for fuels and energy, Curtin University of Technology, Australia. http://www.flyash.info/2005/112ell.pdf (2005). Accessed 04 Sept 2015
7. Mineralogy Database. http://www.webmineral.com. Accessed 04 Sept 2015
8. Jha, B., Singh, D.N.: A review on synthesis, characterization and industrial application of fly ash zeolites. J. Mater. Edu. **33**(1–2), 65–132 (2011)
9. Jha, B., Singh, D.N.: Zeolitization characteristics of a fly ash from wet- and dry- disposal systems. Acta. Geotech. Slov. **2**, 63–71 (2012)
10. Economics of Zeolites. Reports on metals & minerals. ISBN-13: 978-0862149727. Roskill Information Services Ltd. (1988). Accessed 10 September 2014
11. Qiu, L., Murashov, V., White, M.A.: Zeolite 4A: heat capacity and thermodynamic properties. Solid State Sci. **2**, 841–846 (2000)
12. Database of Zeolites Structures. http://www.iza-structure.org. Accessed 22 June 2014
13. Gattardi, G.: Crystal chemistry of natural zeolites. Pure Appl. Chem. **58**, 343–349 (1986)
14. The Mineral Group Zeolite. http://www.minerals-n-more.com. Accessed 10 July 2013
15. Zeolites. http://www.encyclopedia.com. Accessed 20 June 2012

16. Shigemoto, N., Hayashi, H.: Selective formation of Na-X zeolite from coal fly ash by fusion with sodium hydroxide prior to hydrothermal reaction. J. Mater. Sci. **28**, 4781–4786 (1993)
17. Berkgaut, V., Singer, A.: High capacity cation exchanger by hydrothermal zeolitization of coal fly ash. App. Clay. Sci. **10**, 369–378 (1996)
18. Kim, W., Seung, H., Ahn, B.J.: Synthesis of Na-P1 zeolite from coal fly ash. J. Indus. Eng. Chem. **3**(3), 185–190 (1997)
19. Ma, W., Brown, P.W., Komarneni, S.: Characterization and cation exchange properties of zeolites synthesized from fly ashes. J. Mater. Res. **13**(1), 3–7 (1998)
20. Hollman, G.G., Steenbruggen, G., Janssen, J.M.: A two-step process for the synthesis of zeolites from coal fly ash. Fuel **78**, 1225–1230 (1999)
21. Nugteren, H.W., Moreno, N., Sebastia, E., Querol, X.: Determination of the available Si and Al from coal fly ashes under alkaline conditions with the aim of synthesizing zeolites products. International Ash Utilization Symposium. Centre for Applied Energy Research, University of Kentucky. Paper No. 71 (2001)
22. Singh, D.N., Kolay, P.K.: Simulation of ash water interaction and its influence on ash characteristics. Prog. Energ. Combus. Sci. **28**, 267–299 (2002)
23. Chung, S.H., Kim, S.S., Nam, Y.M., Kim, S.M., Lee, B.J.: Synthesis and application of fly ash derived zeolites for pyrolysis of polyolefins. J. Indus. Eng. Chem. **9**(2), 181–187 (2003)
24. Adamczyk, Z., Bialecka, B.: Hydrothermal synthesis of zeolites from polish coal fly ash. Pol. J. Environ. Stud. **14**(6), 713–719 (2005)
25. Inada, M., Eguchi, Y., Enomoto, N., Hojo, J.: Synthesis of zeolite from coal fly ashes with different silica-alumina composition. Fuel **84**, 299–304 (2005)
26. Inada, M., Tsujimoto, H., Eguchi, Y., Enomoto, N., Hojo, J.: Microwave-assisted zeolite synthesis from coal fly ash in hydrothermal Process. Fuel **84**, 1482–1486 (2005)
27. Wu, D., Zhang, B., Yan, L., Kong, H., Wang, X.: Effect of some additives on synthesis of zeolite from coal fly ash. Inter. J. Miner. Process. **80**(2–4), 266–272 (2006)
28. Sulaiman, S.K.: Preparation of synthetic zeolites from coal fly ash. Solid. State. Sci. Technol. **16**(1), 107–113 (2008)
29. Fansuri, H., Pritchard, D., Zhang, D.: Manufacture of low-grade zeolites from fly ash for fertilizer applications. QCAT Technology Transfer Centre, Technology Court, Pullenvale Qld 4069. Australia (2008). Accessed 25 Jan 2009
30. Majchrzak, K.I., Nowak, W.: Application of model-free kinetics to the study of dehydration of fly ash-based zeolites. Thermochim. Acta **413**, 23–29 (2004)
31. Somerset, V.S., Petrik, L.F., White, R., Klinka, M.J., Key, D., Iwuoha, E.I.: Alkaline hydrothermal zeolites synthesized from high SiO_2 and Al_2O_3 co-disposal fly ash filtrates. Fuel **84**, 2324–2329 (2005)
32. Wang, H., Pinnavaia, J.: MFI zeolite with small and uniform intracrystal mesopores. Angewandte. Chemie. Inter. (Ed) **45**, 7603–7606 (2006)
33. Wang, J.: Thermodynamic dehydration and hydration in Natrolite and Analcyme. A thesis presented to the graduate school of the University of Florida in partial fulfillment of the requirement for the degree of Master of Science, University of Florida (2006)
34. Ueda, S., Murata, H., Koizumi, M., Nishimura, H.: Crystallization of mordenite from aqueous solutions. Am. Miner. **65**, 1012–1019 (1980)
35. Kumar, P., Mal, N., Oumi, Y., Yamanaa, K., Sanoc, T.: Mesoporous materials prepared using coal fly ash as the silicon and aluminium source. J. Mater. Chem. **11**, 3285–3290 (2001)
36. Molina, A., Poole, C.: A comparative study using two methods to produce zeolites from fly ash. Miner. Eng. **17**, 167–173 (2004)
37. Querol, X., Alastuey, A., Moreno, N., Alvarez, A.E., Garcı´a-Sa´nchez, A., Cam, J., Ayora, C., Simon, M.: Immobilization of heavy metals in polluted soils by the addition of zeolitic material synthesized from coal fly ash. Chemosphere **62**, 171–180 (2006)
38. Querol, X., Moreno, N., Alastuey, A., Juan, R., Andres, J.M., Lopez-Soler, A., Ayora, C., Medinaceli, A., Valero, A.: Synthesis of high ion exchange zeolites from coal fly ash. Geologica. Acta. **5**(1), 49–57 (2007)

39. Tanaka, H., Fujii, A., Fujimoto, S., Tanaka, Y.: Microwave-assisted two-step process for the synthesis of a single-phase Na-A zeolite from coal fly ash. Adv. Powder Technol. **19**, 83–94 (2008)
40. Davis, M.E., Lobo, R.F.: Zeolite and molecular sieve synthesis. J. Mater. Chem. **4**, 756–768 (1992)
41. Beving, D.E., O'Neill, C.R., Yan, Y.: Hydrophilic and anti-microbial low silica zeolite LTL and high silica zeolite MFI hybrid coatings on aluminium alloys. Micropor. Mesopor. Mater. **108**, 77–85 (2008)
42. Zeng, R., Umana, J.C., Querol, X., Lopez, S.A., Plana, F., Zhuang, X.: Zeolite synthesis from a high Si-Al fly ash from east China. J. Chem. Technol. Biotechnol. **77**, 267–273 (2002)
43. Rayalu, S.S., Bansiwal, A.K., Meshram, S.U., Labhsetwar, N., Devotta, S.: Fly ash based zeolite analogues: versatile materials for energy and environment conservation. Catalysis. Surveys. Asia. **10**(2), 74–88 (2006)
44. Korkuna, O., Leboda, R., Skubiszewska, Z.J., Vrublevs'ka, T., Gun'ko, V.M., Ryczkowski, J.: Structural and physicochemical properties of natural zeolites: clinoptilolite and mordenite. Micropor. Mesopor. Mater. **87**, 243–254 (2006)
45. Jakubinek, M.B., Zhan, B.Z., White, M.A.: Temperature dependent thermal conductivity of powdered zeolite NaX. Micropor. Mesopor. Mater. **103**, 108 (2007). http://dx.doi.org/10.1016/j.micromeso.2007.01.040
46. Mishra, T., Tiwari, S.K.: Studies on sorption properties of zeolite derived from Indian fly ash. J. Hazard. Mater. **B137**, 299–303 (2006)
47. Park, M., Choi, C.L., Lim, W.T., Kim, M.C., Choi, J., Heo, N.H.: Molten-salt method for the synthesis of zeolitic materials I; Zeolite formation in alkaline molten-salt system. Micropor Mesopor Mater **37**, 81–89 (2000)
48. Ojha, K., Pradhan, N.C., Samanta, A.N.: Zeolite from fly ash synthesis and characterization. Bull. Mater. Sci. **27**(6), 555–564 (2004)
49. Tanaka, H., Fujii, A.: Effect of stirring on the dissolution of coal fly ash and synthesis of pure-form Na-A and -X zeolites by two-step process. Adv. Powder Technol. **20**, 473–479 (2009)
50. Kolay, P.K., Singh, D.N.: Physical, chemical, mineralogical and thermal properties of cenospheres from a ash lagoon. Cem. Conc. Res. **31**(4), 539–542 (2001)
51. Querol, X., Moreno, N., Uman, J.C., Juan, R., Hernandez, S., Fernandez, P.C., Ayora, C., Janssen, M., Garcıa, J.M.J., Linares, S.A., Cazorla, A.D.: Application of zeolitic material synthesized from fly ash to the decontamination of waste water and flue gas. J. Chem. Technol. Biotechnol. **77**, 292–298 (2002)
52. Querol, X., Moreno, N., Uman, J.C., Alastuey, A., Hernandez. E., Lopez-Soler, A., Plana, F.: Synthesis of zeolites from coal fly ash: an overview. Inter. J. Coal. Geol. **50**, 413–423 (2002)
53. Lee, S.H., Sakai, Daimon, M., Wang, W.K.: Characterization of fly ash directly collected from electrostatic precipitator. Cem. Conc. Res. **29**, 1791–1797 (1999)
54. Wang, D.J., Tang, Y., Dong, A.G., Zhang, Y.H., Wang, Y.J.: Hollow cancrinite zeolite spheres in situ transformed from fly ash cenosphere. Chin. Chem. Lett. **14**(12), 1299–1302 (2003)
55. Sand, L.B.: Zeolite synthesis and crystallization. Pure Appl. Chem. **52**, 2105–2113 (1980)
56. Rios, C.A.R., Williams, C.D., Roberts, C.L.: A comparative study of two methods for the synthesis of fly ash-based sodium and potassium type zeolites. Fuel **88**, 1403–1416 (2009)
57. Fazio, A.D., Brotzu, P., Ghiara, M.R., Fercia, M.L., Lonis, R., Sau, A.: Hydrothermal treatment at low temperature of Sardinian clinoptilolite bearing ignimbrites for increasing cation exchange capacity. Periodico Di Mineralogia. **77**, 79–91 (2008)
58. Derkowski, A., Franus, W., Waniak-Nowicka, H., Czimerova, A.: Textural properties vs. *CEC* and EGME retention of Na-X zeolite prepared from fly ash at room temperature. Inter. J. Miner. Process. **82**, 57–68 (2006)
59. Mortier, W.J.: Zeolite electronegativity related to physicochemical properties. J. Catalysis. **55**, 138–145 (1978)

60. Fukui, K., Nishimoto, T., Takiguchi, M., Yoshida, H.: Effects of NaOH concentration on zeolite synthesis from fly ash with a hydrothermal treatment method. KONA **24**, 183–191 (2006)
61. Moreno, N., Querol, X., Ayora, C., Alastuey, A., Fernandez, P.C., Janssen, J.M.: Potential environmental applications of pure zeolitic material synthesized from fly ash. J. Environ. Eng. **127**(11), 994–1002 (2001)
62. Jovanovic, V., Dondur, V., Damjonovic, L., Zakrzewska, J., Tomasevic, C.M.: Improved materials for environmental application: surfactant-modified zeolites. Mater. Sci. Forum **518**, 223–228 (2006)

Chapter 3
Conventional Methods for Synthesis of Fly Ash Zeolites

Abstract Though, naturally occurring and chemically synthesized (pure grade) zeolites have been used for various industrial applications in the past, their increasing demand for several novel applications (viz., as adsorbent or absorbent for waste water decontamination, soil remediation as fertilizers, aqua-culture purification, etc.) warrants their enhanced production. With this in view, several researchers have attempted to synthesize zeolites from the fly ash, an abundantly available industrial by-product, as described in this chapter. Furthermore, different methods employed for synthesis of fly ash zeolites, the mechanism of zeolites formation and potential fields of their applications have also been included herein.

Keywords Synthesis of zeolites · Fly ash · Alkali activation · Hydrothermal · Fusion · Microwave irradiation

3.1 Methods of Synthesis of Zeolites

A brief account of the research pertaining to the methods employed for synthesis of fly ash zeolites, FAZ, (viz., conventional hydrothermal, fusion, microwave assisted hydrothermal, molten salts and other two stage hydrothermal) is presented in Table 3.1. In addition, the effect(s) of variation in reaction parameters (listed in Table 3.1) on the end product (read the type of zeolite(s) formed) and their characteristic properties (as discussed in Table 2.11) have also been discussed herein. By going through this section, attempts have been made to understand the state of the art available and devise a new methodology to investigate and analyze the various outcomes of the experiments and to explore new ideas in the field of utilization of solid waste (i.e., fly ash) in the coming generation of research.

B. Jha and D.N. Singh, *Fly Ash Zeolites*, Advanced Structured Materials 78,
DOI 10.1007/978-981-10-1404-8_3

Table 3.1 A summary of the methods employed for synthesis of fly ash

Method	L/S	Reagent	Temp. (°C)	*T* (h)	Zeolite	Remarks	References
Conventional hydrothermal	8	NaOH, KOH, Na_2CO_3	90–150	24–96	Chabazite, Na-P1, Phillipsite, Sodalite, zeolite KH, 4A, A, P, Zeolite X, Y	Low yield, low purity, structural-heterogeneity	[1–5]
Microwave assisted hydrothermal	8	NaOH	100	0.25–2	Na-P1	Synthesis time very less	[6–8]
Fusion and Hydrothermal	10	NaOH/or Sodium aluminate	500–650	1–2	Faujasite Na-A, Na-X or zeolite X	More yield	[1, 9–14]
		Water	90–100	6			
Molten salt	–	KOH KNO_3 NaOH $NaNO_3$ NH_4F NH_4NO_3	350	3–6	Sodalite	No addition of water, irregular morphology	[15–17]
				24	Cancrinite		

T Activation time, *Temp.* temperature, *L/S* liquid to solid ratio
– not any

3.1.1 Conventional Hydrothermal Method

The conventional hydrothermal method (either open or closed reflux systems) for alkali activation of fly ash has been extensively utilized by previous researchers [1, 5]. A typical experimental set up employed for the closed reflux system is similar to an autoclave [18], where both pressure and temperature can be varied as per the desired experimental conditions.

In order to investigate the effectiveness of different types of alkalis, for activation of the fly ash, researchers have, mostly, adopted NaOH or KOH aqueous solutions by varying the temperature between 80 and 200 °C and activation period between 3 and 48 h. 13 different types of fly ash zeolites have been synthesized by employing this method [3, 5, 6, 19–28].

In context to the characterization of the synthesized zeolites, it has been reported that the quantity of zeolitic minerals present in the end product has been found to vary widely ($\sim$20–65 % by weight of fly ash), mainly depending on the various parameters (viz., type of fly ash, alkali concentration, temperature, time and liquid to solid ratio) involved in the synthesis process. A typical flow chart is depicted in Fig. 3.1 to highlight the different procedural steps required for this method.

Kim et al. [19] have synthesized Na-P1 zeolite from coal fly ash by using aqueous solution of 2–3 M NaOH at 100 to 150 °C. It has been reported that Faujasite and Hydroxy-sodalite are formed when NaOH solution concentrations are maintained at 4 and 5 M, respectively. It has been observed that all Mullite, Quartz and Hematite disappeared after hydrothermal treatment at 160 °C. The final synthesis product containing zeolites, Analcime, with Na-P1, and Cancrinite with Na-P1, can be found corresponding to alkali concentration of 2 and 4 M, respectively. It has also been reported that Cancrinite increases if the NaOH concentration is 5 M or more. The authors have opined that Na-P1 zeolite having *CEC* value of 215 meq./100 g would exhibit strong affinity towards Pb and Sr.

Hollman et al. [29] have employed a two-step process for the synthesis of pure zeolites from coal fly ash with better yield. It has been reported that 50–85 g of pure

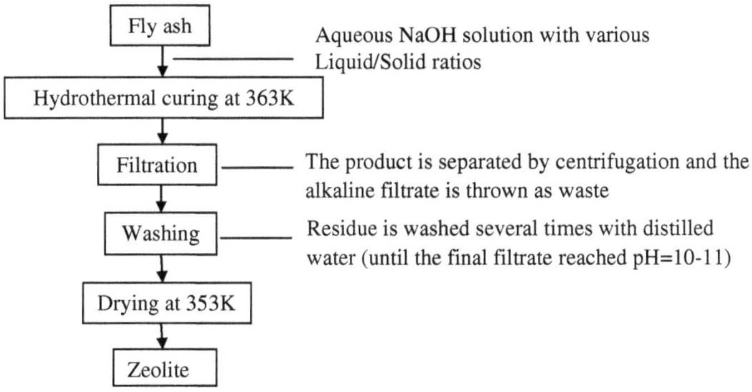

Fig. 3.1 Flowchart of conventional hydrothermal synthesis process

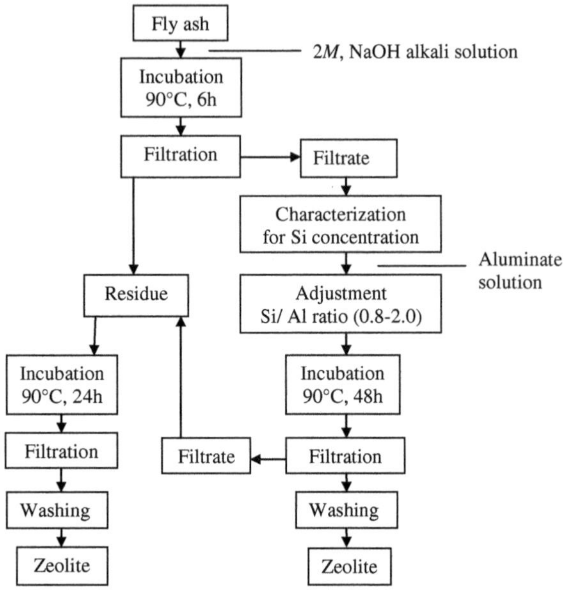

Fig. 3.2 Flowchart for two step process for synthesis of fly ash zeolites

zeolites Na-P1, Na-X, Na-A and Sodalite can be synthesized per kg of coal fly ash sample. The measured cation exchange capacities of such pure zeolites have been reported to range from 360–430 meq./100 g unlike 200–250 meq./100 g for zeolites containing residual fly ash as impurities, synthesized by a conventional single step process of hydrothermal activation. The flow chart adopted for the two stage synthesis process is presented in Fig. 3.2.

Kolay et al. [30] have synthesized zeolites from the lagoon ash. It has been reported that physical and chemical properties of the fly ash get changed due to the reaction between SiO_2 and Al_2O_3 with Na_2O and K_2O present in it, under wet conditions. This results in formation of Na-P1 zeolites over a period of 12 h, at 100 °C. It has been concluded that alkali activation of the fly ash causes a decrease in its silica content because of etching, which leads to the creation of some fines, and agglomeration of the fines to form coarser particles. It has been reported that the activation of lagoon ash with KOH does not result into its zeolitization.

Nugteren et al. [17] have investigated the available Si and Al after alkaline leaching of coal fly ash with the aim of synthesizing pure zeolitic products. It has also been opined that the zeolites also get formed on the surface of ash particles, during the process, which limits the extraction of Si. The Al content in the extract (i.e., the solution enriched in Si and Al, obtained after the treatment, popularly termed as filtrate or supernatant) is reported to be almost negligible which gets consumed for synthesis of zeolites. Hence, with an intention to delay crystallization of zeolites, the extraction of both Si and Al from the RFA has been tried at lower temperature (=50 °C), with L/S ratio of 20, 2.5 M NaOH solution and reaction time of 72 h. It has been reported that this treatment increases extraction of Si upto

280 g/kg of ash, with few zeolites that consume some Si and Al in the extract. Hence, Al complexants (viz., gluconic acid, tartaric acid, Na-gluconate and sorbitol) have been added to the reactant NaOH solution to restrict precipitation of zeolites and enhance Si and Al concentration in the final extract.

Fernandez et al. [31] have synthesized zeolites by reflux heating of stirred mixture of fly ash and 2.5 M NaOH and/or KOH solution, on an electrically heated plate at 100 °C for 24 or 48 h by maintaining solution to fly ash ratio (L/S) as 0.005 L/g. The potential reuse of waste alkaline solution has been evaluated for activation of a fresh fly ash sample. It has been observed that there is no change in the *CEC* value of zeolitic product obtained, which is an anomaly and should be researched further. It has also been reported that the zeolitic yield has been up to 38 % (viz., Na-P1 and Analcime) and 34 % (viz., Chabazite) for NaOH and KOH activations, respectively. The mineral phase transition has been attributed to the activation of majority of amorphous glass with NaOH, whereas Mullite and Quartz can get dissolved under strong alkaline conditions. It has been demonstrated that the products obtained after NaOH activation can get contaminated with more leachates and trace elements like Cd and As, as compared to KOH activation from fly ash.

Kolay and Singh [4] have investigated the alkali activation of lagoon ash and have confirmed the formation of zeolites Na-P1 and Hydroxy-sodalite because of its silica dissolution over a period of time (viz., 12–48 h). The activated ash samples have been characterized to study the effect of zeolitization of ash on their physical (viz., particle size, specific gravity and surface area), chemical (viz., chemical composition and cation exchange capacity), mineralogical and morphological properties. They have optimized the alkali concentration (e.g. 1 M NaOH) and reaction period of 12 and 24 h of hydrothermal activation for synthesis of zeolites Na-P1 and Hydroxy-sodalite, respectively. The net increase in specific gravity (≈2–11.8 %), specific surface area (≈0.72–142.51 %) and cation exchange capacity (≈0–150 %), has been reported owing to alkali activation for varying time and NaOH concentrations.

Murayama et al. [3] have explained the mechanism of zeolite synthesis from coal fly ash by its hydrothermal reaction with alkali. They have observed that alkaline medium type (viz., NaOH, Na_2CO_3 and KOH) affects the mechanism of crystallization of zeolites. The authors have employed synthesis matrix bearing a specific solid/slurry ratio (i.e., 100 g/400 cm^3), with the slurry being an aqueous mixture of two different alkalis (viz., NaOH and Na_2CO_3, NaOH and KOH and Na_2CO_3 and KOH) to investigate the effect of the presence of different cations and/or anions on the alkali activation of fly ash. It has been reported that zeolites, P and Chabazite, are the main crystals present in the synthesized product. The OH^- in the alkali solution remarkably contributes to the dissolution of Si^{4+} and Al^{3+} from coal fly ash, whereas Na^+ makes a contribution to the crystallization of zeolite P, which has the tendency to capture K^+ in the cation exchange process.

Singh and Kolay [32] have investigated the process simulation of ash-water interaction and its influence on various geotechnical properties (viz., compaction, consolidation, hydraulic conductivity and shear strength) of lagoon ash as a fill material for its bulk utilization. They have demonstrated that the activation of ash with NaOH has been essential for formation of zeolites [33]. The authors have concluded that the activation of lagoon ash with KOH has been ineffective for synthesis of zeolites.

Chung et al. [23] have investigated the effects of NaOH concentration and reaction time on the mineralogical structure of zeolites synthesized by hydrothermal treatment of fly ash with NaOH solution. It has been reported that the effect of these zeolites on the pyrolysis behavior of low density polyethylene and polypropylene can be delineated in terms of recovery and distribution of boiling points of liquid pyrolysis products. It appears that the fly ash derived zeolites are not effective for low density polyethylene (LDPE) pyrolysis but slightly effective for polypropylene pyrolysis.

Fukui et al. [33] have investigated the effects of NaOH concentration on the crystal structure and the rate of reaction of the synthesized zeolite from fly ash with a hydrothermal treatment method. They have reported that fly ash or the mixture of fly ash and silica powder results in an increase in the reaction rates with the increase of NaOH concentration due to increase of the dissolution rate of silicate ion and aluminate ion. It has been clarified that the NaOH concentration also affects the crystal structure of synthesized zeolites. It has been concluded that the proportion of Phillipsite continues to be lower than the increasing proportion of Hydroxy-sodalite in the product with the increase in the concentration of NaOH.

Adamczyk and Bialecka [5] have established the process of hydrothermal synthesis of zeolites and optimized the conditions of synthesis to get as much quantity of zeolite as possible in the shortest activation time. The process of preliminary activation has been carried out at room temperature with NaOH solution. In addition, further activation has been carried out at variable temperatures (viz., 80, 120, 140, 150, 170, 180, 200 and 320 °C), and at 6 h reaction time, in a 1 l PROLABO autoclave fitted with a stirring electromagnetic system at activation pressure equal to the pressure of the steam generated in the course of heating the suspension. It has been reported that there has been an increase in the content of Na-P1 zeolite as compared to Analcime from 120 to 170 °C of activation temperature, thereafter its content gets reduced at 180 °C and again increased beyond this temperature up to 320 °C along with formation of more Analcime. Diffraction patterns of the material synthesized at 120 °C demonstrate the presence of Mullite and Hematite reflections.

Inada et al. [6] have synthesized zeolites employing different coal fly ash sources, bearing different silica-alumina compositions by hydrothermal alkaline treatment. It has been observed that zeolites Na-P1 and/or Hydroxy-sodalite appear after the treatment at 100 °C in aqueous NaOH solution for 24 h. It has been demonstrated that zeolite Na-P1 can be synthesized predominantly from silica-rich fly ash at a low NaOH concentration with maximum yield at 2 M NaOH. They have opined that the presence of Hydroxy-sodalite can be found by employing 4 M NaOH concentration, whereas zeolite Na-P1 can fully disappear at 5 M NaOH. The cation exchange capacity of the product with a large content of zeolite Na-P1 has been found to reach a value of 300 meq./100 g. It has been concluded that silica addition effectively enhances the formation of zeolite Na-P1, even at a high NaOH concentration. They have elaborated these results on basis of a formation mechanism of zeolite from coal fly ash through dissolution-precipitation process.

Fernandez et al. [31] have demonstrated the role played by the reactive alumina content in alkaline activation of fly ashes. It has been confirmed that the amount of the reactive alumina plays an important role in the formation of aluminosilicate gel in a strong alkaline concentration of 8 M NaOH, at 85 °C, as the curing time varies from 2 h to 7 days. Usually, the fly ash which has reactive SiO_2 and Al_2O_3 contents in the ratio below 2 is considered to be most appropriate for alkaline activation. Authors have confirmed that alkaline aluminosilicate gelis formed as a major reaction product, regardless of the composition of the ash. It has also been observed that the gel formed in short reaction time, constitute of alumina rich phase in which a Si tetrahedron is surrounded by four Al tetrahedron.

Tanaka et al. [7] have devised two step processes for synthesis of a single phase Na-A zeolite from coal fly ash by dialysis and using amorphous aluminosilicate glass content of the raw ash. They have identified the mineral phases present in fly ash and results are depicted in Table 3.2 [7]. The first step of hydrothermal activation has been employing the mixture of fly ash and NaOH at 85 °C for 24 h, followed by the second step of adjustment of SiO_2/Al_2O_3 molar ratio (viz. 0.9–4.3) of residual solution by addition of solution of a mixture of NaOH and $NaAlO_2$. The dissolution of the amorphous aluminosilicate in the solution has been reported to occur which can finally yield a white precipitate (i.e. identified as Na-A zeolite) at SiO_2/Al_2O_3 molar ratio of 0.9. The precipitation of Na-X zeolite has been demonstrated for $SiO_2/Al_2O_3 \geq 1.7$.

Wu et al. [34] have studied the effect of additives on synthesis of zeolite from coal fly ash due to hydrothermal conversion. The authors have demonstrated the effect of addition of sodium halide, Na_2SiO_4 and $Al(OH)_3$, prior to synthesis process for the adjustment of the Si/Al ratio of the reaction matrix, on waste solution and the *CEC* of final products produced, after zeolitization of the fly ash. The addition of NaCl and NaF has been found to ameliorate the crystallinity and *CEC* of the synthesized zeolite. However, NaF has been reported to be a better option for the improved effects. It has been concluded that Na^+ enhances the crystallization of zeolites, while F^- favors the dissolution of the fly ash. It has been observed that the waste solutions contain large amount of Si and little Al, due to the formation of a zeolite, Na-P1 (as per the Joint Committee of Powder Diffraction Standard, JCPDS,

Table 3.2 Content of major mineral phases and their oxides present in raw fly ash: from two different sources

Major oxides	Minerals	Phase	Content (% by wt.)	
			I[a]	II[a]
SiO_2	Mullite, quartz	Crystalline	15.8	14.8
	Glass	Amorphous	53.7	49.4
Al_2O_3	Mullite	Crystalline	12.2	18.8
	Glass	Amorphous	5.0	4.5

[a]Source of fly ash

code of 39-0219). It has also been concluded that the waste solutions produced following the zeolitization of fly ash can be reused as an alkali source for activating the fly ash and hence 'zero-emission' of the waste solution in the synthesis of zeolite from the fly ash is possible.

Criado et al. [35] have studied the effect of the SiO_2/Na_2O ratio on the alkali activation of fly ash on the micro structural and mechanical development of the cementitious materials produced as the final yield. The fly ash has been activated by four different alkaline solutions with different soluble silica contents. It has been clarified that the primary reaction product has been a sodium aluminosilicate gel, while different types of zeolites appeared as secondary products, in minority. The percentage and composition of these reaction products have been found to depend on both the soluble silica content present in the activating solutions and the thermal curing time. In addition, the amount of gel formed, has been found to have a decisive effect on the mechanical strength of the material. The inclusion of a greater amount of soluble silica in the environment has been opined to retard the zeolite formation, mainly due to the higher degree of polymerization of available silica.

Derkowski et al. [36] have established the variation of the textural properties versus *CEC* and ethylene glycol monoethyl ether (EGME) retention of Na-X zeolite prepared by storing fly ash in NaOH solution for 7–20 months at room temperature. They have demonstrated that the action of $(OH)^-$ ion on ash and subsequent dissolution led to the formation of Na-X zeolite yield up to 50 %. They have opined that the dissolution process can be accelerated by employing solid to liquid (S/L) ratio of 33–67 g/l. It has been noticed that the Si/Al ratio of synthesized products ranges from 1.14 to 1.45, which ultimately affects their micro-pore volume, surface area, EGME retention and the cation exchange capacity (*CEC*). It has been opined that pre-heating of the samples up to 250 °C does not allow EGME molecules to enter into the micro-pores due to insufficient dehydration. Whereas, pre-heating at 400 °C caused their adsorption on all available surfaces. They have confirmed that the *CEC* values of samples to be around 120 meq./100 g, whereas N_2-BET surface area get enhanced from 15.3 to 344 m^2/g because of the development of micro-porous texture of fly ash particles.

Fansuri et al. [37] have reported the process of manufacturing low grade zeolites from the fly ash, mainly for their application as fertilizer. A fertilizer grade zeolite (viz., a mixture of Analcime, zeolites P and P1) has been produced by a 4 h long hydrothermal reaction. It has been reported that spinach and wheat exhibit better growth due to up to 5 % addition of these zeolites in a yellow gravely loamy sand. It has been observed that K-zeolite is responsible for better growth of these commodities as compared to Na-zeolites.

Sulaiman [38] has synthesized fly ash zeolites (FAZ) from coal fly ash. It has been reported that the hydrothermal reaction of magnetically pre-treated fly ash can produce stable zeolites P by initiating its activation in aqueous NaOH solution at 70 °C for 48 h. The crystallinity of the zeolites has been found to increase with the increase in temperature and reaction time. It has been concluded that the mineral transformation takes place due to dissolution of aluminosilicate glass and formation of zeolites P.

Wałek et al. [39] have demonstrated that low solid to liquid (S/L) ratio (i.e., up to 10 g/l) can cause a significant increase in fly dissolution (i.e., up to 85 %) during first 4 h of reaction with aqueous solution of NaOH, whereas the dissolution of fly ash ingredients get limited for higher S/L ratios (≈50–150 g/l). It has been reported that zeolite crystals are formed on the surface of partially activated fly ash particles for matrix bearing higher S/L ratio, which can restrict further dissolution. The authors have confirmed the formation of zeolites Na-P1, Analcime and/or Hydroxy-sodalite, which bear the degree of crystallization up to 80 %, for 2 M NaOH solution and at temperatures of 104 and 80 °C during the stages of dissolution and crystallization, respectively.

3.1.2 Microwave Assisted Hydrothermal Method

In order to reduce the total time of synthesis (≈30 min), it has been reported to employ two stage synthesis processes, an initial microwave heating for around 30 min. followed by the conventional synthesis method [8, 40, 41]. Although, the purity of zeolite has been of much concern, the final yield used to contain unreacted (read residual) fly ash, which can affect the characteristics of synthesized zeolite. In fact, the presence of considerable quantity of unreacted fly ash residue in the final yield can be attributed to the factors, which are being discussed in the next chapter: "The mechanism of zeolite formation".

Inada et al. [41] have investigated the microwave assisted zeolite synthesis process by employing fly ash in its hydrothermal activation. It has been demonstrated that zeolite Na-P1 can be formed after the conventional treatment using oil bath, but no zeolites appear in the product obtained by microwave heating for long period. With this in view, microwave irradiation has been limited to early-stage (i.e., about first 15 min.) of the hydrothermal activation to initiate zeolitization process. They have also demonstrated that the dissolution of SiO_2 and Al_2O_3 from coal fly ash can be accelerated by fast heating up of the mixture of fly ash and 2 M NaOH aqueous solution in microwave technique. It has also been opined that active water is generated by breaking of hydrogen bonds which attacks Si-O and Al-O bonds; this in turn enhances the dissolution of the glass phase on surface of ash particles.

Tanaka et al. [42] have studied the microwave assisted two-step process for the synthesis of a single phase Na-A zeolite from coal fly ash. It has been reported that the amorphous constituents of fly ash gets dissolved very fast into the aqueous NaOH solution during its microwave irradiation of refluxing heating system for 1 h. They have observed that white precipitates of Na-X is formed after heating the mixture of treated fly ash and NaOH-$NaAlO_2$ solution in air oven at 358 K for 24 h. They have opined that Hydroxy-sodalite also can be formed corresponding to SiO_2/Al_2O_3 ratio $\geq$ 1.0 whereas; Na-X and Na-A is formed for SiO_2/Al_2O_3 ratio $\geq$ 4.0 and 0.4–0.8, respectively.

Fukui et al. [8] have utilized NaCl for synthesis of Phillipsite from fly ash by hydrothermal treatment with microwave heating. It has been reported that the crystallinity of Phillipsite can be increased by addition of NaCl in the hydrothermal reaction matrix with low concentration of NaOH, whereas no effects have been reported in case of high concentration of NaOH. However, microwave heating has been a favorable tool for the generation of Hydroxy-sodalite. It has been demonstrated that the substitution of NaCl by NaOH can reduce the rate of dissolution of aluminate and silicate ions from fly ash. As such, the rate of generation of aluminosilicate gel can be enhanced by employing microwave heating of the solution.

3.1.3 Fusion and Hydrothermal Method

In order to augment the quantity and quality of final synthesis yield as obtained from conventional hydrothermal activation, another modified method has been introduced which utilizes two different steps, an initial high temperature fusion of fly ash-alkali mixture, prior to employing the final stage of hydrothermal activation of the fused product. The main variables have been fusion temperature and time, alkali type and its concentration and crystallization time in hydrothermal synthesis process, which can affect the quality and yield of final product. As such, it has been confirmed that the final yield can be quantified to exhibit zeolitic conversion up to 62 % together with by production of alkaline waste solution which can become a threat to the environment after disposal. A flowchart of the synthesis process is depicted in Fig. 3.3 [1, 2, 9, 10, 12, 43, 44].

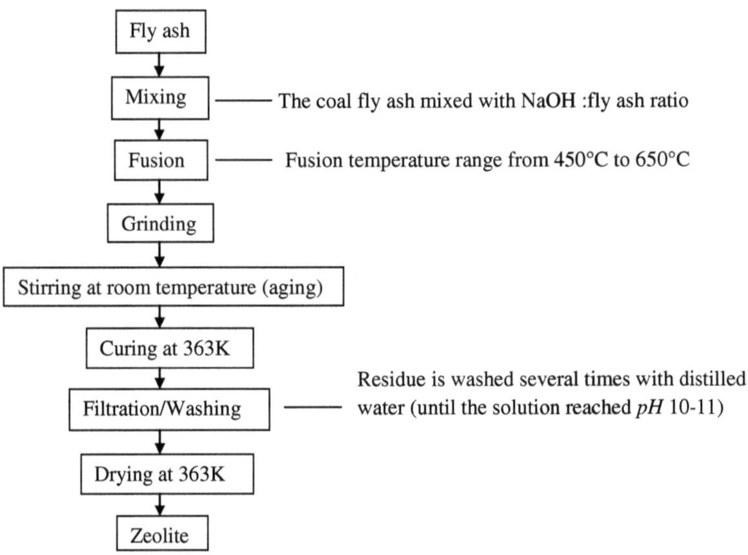

Fig. 3.3 Flowchart of fusion prior to hydrothermal synthesis process

Shigemoto et al. [9] have reported about selective formation of Na-X zeolite from a mixture of coal fly ash procured from two different sources. It has been reported that most of the fly ash particles got converted into silicates and aluminates of sodium during 1 h of fusion at 773 K, whereas the hydrothermal activation for 6 h of the fused product, crystallized into Na-X, Na-A, Na-P1 and Hydroxy-sodalite zeolites in different proportions based on Na/Fly ash ratio (i.e., 1.2–1.8) employed and Al content of fly ash. It has been opined that fly ash enriched with aluminium, can be crystallized to zeolite Na-A in place of Na-X.

Berkgaut et al. [10] have synthesized a high capacity cation exchange material by hydrothermal zeolitization of coal fly ash. They have reported that the ash partially converts to zeolite P during hydrothermal treatment in NaOH medium at 100 °C while the same product gets converted to Sodalite after 12 h ageing. It has been observed that their XRD patterns represent the decrease in peak heights of Quartz due to its dissolution while Mullite peaks remain unchanged after hydrothermal treatment. The fusion treatment of the same sample has demonstrated the decomposition of Mullite. They have concluded that alkaline fusion prior to hydrothermal treatment at 170–80 °C, results in aluminosilicate (i.e., approx. $Na_{15}Si_4Al_3O_{20}$) as a major product leading to 100 % conversion of ash to zeolites.

Grutzeckand Siemer [45] have investigated the synthesis of zeolites from a mixture of class-F fly ash and slurry of sodium aluminate prepared in proportion of 3:1 (i.e., Na:Al). They have reported that a class-F fly ash reacts with the slurry to synthesize zeolites from a highly alkaline waste stream. The reaction has been studied as a function of mixture (i.e., fly ash: slurry) composition (3:2, 1:1, and 1:2), time (1, 3, and 7 days), and temperature (80, 130, and 180 °C). The X-ray diffraction analysis of the products has indicated that the reaction between fly ash and sodium aluminate can result final yield as zeolite A, Na-P1, and Hydroxy-sodalite at 80, 130 and 180 °C, respectively. It has been clarified that the bulk of the sodium has been incorporated into the zeolitic phases.

Rayalu et al. [1] have synthesized fly ash zeolites (FAZ) by fusing fly ash and sodium hydroxide to produce zeolites Y subjected to subsequent hydrothermal activation. The crystallization of zeolite Y has been reported to increase with NaOH concentration (optimum being 22–24 g). However, further increase in NaOH concentration from 32–40 g has been found to lead to transformation of zeolite Y phase to Hydroxy-sodalite. The SiO_2/Al_2O_3 ratio of 3.3–3.5 in the fly ash has been reported to favor formation of zeolite Y. The *CEC* of FAZ-Y having cubic crystals has been found to range from 400 to 450 meq./100 g. It has been demonstrated that the surface area of FAZ is of the order of about 500–600 m^2/g. The point to be noted here is that zeolites X and A can be synthesized by addition of sodium aluminate in the synthesis matrix to increase the Al_2O_3 content and modify the actual SiO_2/Al_2O_3 ratio.

Kumar et al. [2] have synthesized meso-porous, MCM-41 type zeolitic material from supernatant obtained after filtration of aqueous solution (L/S = 4) of fused mixture of NaOH and fly ash, which has Brunauer-Emmett-Teller (BET) surface area (SSA_{BET}) and *CEC* around 4.5 m^2/100 g and 0.8 meq./100 g, respectively. They have concluded that most of the Si and Al components in fly ash can be effectively extracted into supernatant to synthesize a pure zeolite (with Si/Al = 13.8 and SSA_{BET} = 842 m^2/g) by employing subsequent hydrothermal activation. It has been demonstrated that fusion can play an important role in enhancing the zeolitization of fly ash in hydrothermal conditions. In addition, SBA-15 (with Si/Al = 653 and SSA_{BET} = 483 m^2/g), a type of siliceous materials can also be synthesized from coal fly ash.

Molina and Poole [46] have synthesized zeolites from a magnetically pre-treated fly ash (PFA) by employing hydrothermal and fusion methods and carried out their comparative study. They have demonstrated that the pre-treatment of fly ash can lead to increase its Si and Al content together with a huge decrease (up to 65 %) in the iron like impurities. It has been clarified that optimized condition of temperature, time and proportion of NaOH:PFA:H_2O, can lead to the synthesis of zeolites X and Hydroxy-sodalite corresponding to increase in NaOH content in fusion method, whereas the presence of zeolite X and A together with original crystalline phases of Quartz and Mullite, have been confirmed in hydrothermal method. It has been concluded that the fusion method can yield better results in shorter time. As such, the *CEC* of end products has been reported to increase by 40 and 50 % by employing hydrothermal (up to 100 meq./100 g) and fusion method (up to 200 meq./100 g), respectively, as compared to that of raw fly ash (≈4 meq./100 g).

Ojha et al. [43] have synthesized zeolites from fly ash by fusion. The authors have worked for calcined fly ash of fineness more than 80 μ (the mesh size). The fly ash has been pre-treated with HCl to dealuminate and to remove its iron contents to increase the reactivity. The pre-treated fly ash and sodium hydroxide mixture in ratio of 1:1.5 has been analyzed after fusion at temperature of 500–650 °C for 1 h. The fusion product containing sodium silicates and sodium aluminates has been converted to slurry with water in 1:10 (ash:water) ratio and observed for several hours under ageing. It has been demonstrated that the resulting mixture kept for crystallization at 90 °C for 6 h, finally can exhibit the presence of Na-X type zeolites with high crystallinity and excellent performance as cation exchanger.

Rayalu et al. [12] have estimated the crystallinity of fly ash zeolites-A using XRD and infrared (IR) spectroscopy for identification, quantification and their framework structure. Zeolite A has been synthesized by fusion of mixtures of fly ash and sodium hydroxide in 1:1.2 ratios at temperatures from 550 to 600 °C for 1–1.5 h of heating time. It has been reported that powder XRD analysis was employed to monitor the formation of zeolite-A. The infrared, IR, technique has been proposed for monitoring crystallinityof end product after synthesis. Based on the characterization results, it has been opined that the unreacted fly ash associated with the zeolite-A as impurities in the final yield, have negligible influence on its application as an adsorbent or cation exchanger. As such, the crystallinity of end product as per interpretation of XRD peaks of commercial zeolite-A have been reported to match closely with the crystallinity interpreted from IR spectrum of the mineral.

Rungsuk et al. [44] have studied the effect of NaOH concentration, temperature and time of fusion in the synthesis of zeolite-X. It has been demonstrated that the *CEC* of the zeolites can get reduced from its maximum of 154.7 meq./100 g corresponding to NaOH/CFA ratio beyond 2.25. They have optimized the fusion temperature and time to 450 °C and 2 h, respectively, which can result maximum *CEC* as high as 220 meq./100 g based on attributes such as the complete reaction and damage to zeolite crystals owing to longer reaction time and higher temperature.

Yao et al. [47] have synthesized zeolite Li-ABW from fly ash by alkaline fusion followed by hydrothermal treatment in $LiOH \cdot H_2O$ medium. It has been reported the crystallinity of zeolites is as high as 97.8 % corresponding to 3 M $LiOH \cdot H_2O$ concentration, hydrothermal temperature range 80–180 °C and aging time, 2–16 h. It has been found that the soluble species in the supernatant produce different zeolitic crystals by transformation after re-crystallization. They have confirmed the presence of rod shaped stable crystals which consist of meso-pores by employing characterization tools such as XRD, FT-IR, SEM, TGA-DTA and N_2 adsorption isotherms.

3.1.4 Molten Salt Method

In view of the production of alkaline supernatant/waste solution after hydrothermal and fusion methods, a molten salt method has been demonstrated by previous researchers for synthesis of zeolites by employing a dry conversion technique without use of water. In tune with this, mixture of chemicals/salts and fly ash has been reported to yield zeolites by reaction of salts in molten condition for a period up to 3 days. A flow chart of the synthesis process is depicted in Fig. 3.4 [15, 16].

Park et al. [15, 16] have established the molten-salt method for the synthesis of zeolitic materials without any addition of water. The zeolitization of fly ash has been demonstrated by employing various mixtures of NaOH, KOH, or NH_4F as mineralizers and $NaNO_3$, KNO_3, or NH_4NO_3 as solvent instead of demineralized water in hydrothermal method. It has been reported that zeolites like Hydroxy-sodalite and Cancrinite can get crystallized from mixture of NaOH and

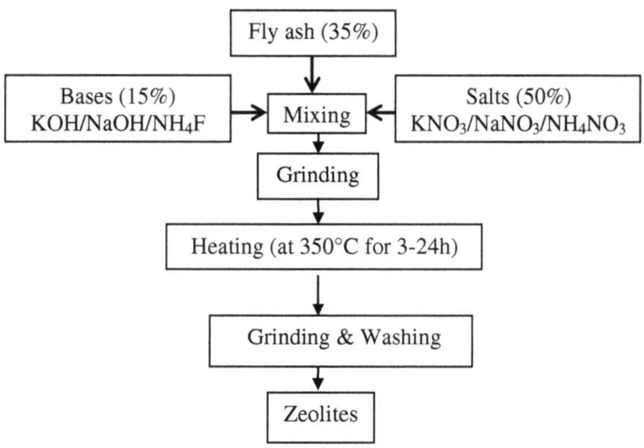

Fig. 3.4 Flowchart of molten salt method for synthesis of zeolites

$NaNO_3$ for activation times of 3–6 and 24 h, respectively; at 350 ± 5 °C. It has been observed that the replacement of $NaNO_3$ byKNO_3 in the mixture stabilizes Cancrinite structure. With this in view, it has been concluded that the Na^+ has higher ability of zeolitization than K^+ cations, probably due to the higher reactivity of smaller cations with raw materials. However, it has also been realized that low temperature and insufficient contact of NaOH with particles of fly ash can lead to incomplete zeolitization of the ash [11].

3.2 Critical Appraisal

A critical review of the literature available on synthesis of fly ash zeolites reveals that the hydrothermal activation of the fly ash with alkaline material, confirms zeolitization of the ash, although; the effect of variation in NaOH concentration over physical properties (viz., particle size, specific gravity, pore size, void volume and surface area), has not been studied in detail and hence proper trends of inter-dependence between these parameters is not available. Moreover, the liquid by-product of the hydrothermal activation method is a threat to environmental cleanup. The last supernatant after the hydrothermal synthesis process has to be thrown out as unused waste in the environment, which can be harmful and a threat to the environmental safety. With this in view, modifications like two stage synthesis should be tried by recycling/reuse of the first filtrate/supernatant for further reduction of the alkalinity, separation of dissolution and crystallization process to avoid the presence of nonreactive and residual glass in the final product. In line with this objective, and to produce the last supernatant with least alkalinity and minimum concentration of metal ions incorporated or extracted during synthesis process, multi stage recycling is the matter of concern, as much attention has not been paid in this context.

As an alternative, the use of water in the synthesis process can be eliminated to avoid the production of liquid waste as reaction by-product. However, the complete conversion of fly ash to zeolites has not been possible due to insufficient contact of Na^+, in the molten stage, with the surface of fly ash particles. In addition, the yield of this molten salts reaction process has been low and this combined with the generation of irregular product morphology has led to the processes which have not been explored much.

As a step ahead, researchers have further modified the synthesis process to maximize the consumption of alkalinity in zeolitization and ash dissolution by resorting to 'fusion method' to produce zeolites, in less time and that too with higher yield and better properties. Based on the earlier studies it has been observed that not much has been explored about the need of preparation of homogenous mixtures of synthesis matrix [6]. The matrix connotes to a material which consists of heterogeneous composition of several oxides such as glass in the fly ash, NaOH and water (i.e., the most important parameter) of NaOH and fly ash and its effect on the ingredients of fusion products [10]. There are possibilities of loss of concentration of molten alkaline material, insufficient contact of alkali cations and anions with the ash particles and thus incomplete fusion reaction to produce sodium and aluminum silicates. Furthermore, the formation of a precursor gel for zeolitization initiates during hydrothermal activation of fused product on addition of water, the final alkali concentration in aqueous phase, maintaining adequate liquid to solid ratio of the mixture subjected to alkali activation as discussed above. However, loss of weight during fusion and its effect on the properties of the final product, and their superiority for industrial applications, remain un-quantified.

With this in view, the change in the mineral phase of fly ash, during its varying degree of thermal decomposition and hence alkali-fly ash interaction in the dry stage of fusion and wet stage of hydrothermal activation needs to be optimized to better understand the superiority of both the methods with a special focus to the mechanism of zeolitization of the fly ash, discussed in detail, in the next chapter.

Furthermore, the time of activation has been reduced significantly by employing microwave assisted heating in hydrothermal method. Although, the alkali dissolution of fly ash can be accelerated by fast heating and attack of active water at molecular level, however, it has been found that the end product is of low yield, which has not been explored further. In addition, no efforts have been made for understanding the effects of the physical variation of the fly ash particles on the quality of final products. With this in view, an optimization of the time of microwave heating of aqueous matrix of different L/S ratio, alkali concentration as well as the type of fly ash collected from various sources (viz., bottom ash, hopper ash, ash from electrostatic precipitator), needs to be carried out.

With all these challenges ahead, thorough investigations are required to establish the effects of reactivation or recycling of the activated fly ash residue with the supernatants over its physico-chemical, mineralogical and morphological characteristics and their variations in the synthesized products. In order to investigate the zeolitization potential of a raw fly ash sample and its activated residue produced after each stage of activation, it is required to characterize the final product to correlate the decrease in the quantity of impurities or the inactivated fly ash particles in the final product with the different stages of reactivation. In addition, the variation in the concentration of the silica and aluminum in the solution can be correlated with the degree of zeolitization of fly ash. With this in view, the thorough investigation for all these possibilities discussed above, needs to be the matter of concern for the future researchers.

However, it has been opined by that the highest release of SiO_2 in the solution indicates the lowest formation of zeolites, although; the slowing down of dissolution can lead to higher zeolitization. It has been reported that the dissolution of Al is found to be below average at short reaction times for Al rich phases (i.e., Mullite) and/or above average for dissolution of Si rich phases (i.e., high Si-glass). As such, the increase in Al content of the solution corresponding to longer reaction period has been demonstrated to result in the precipitation of zeolites. In contrast to this, the optimization of the synthesis methods and activation condition (viz., alkali type and concentration, L/S, temperature and pressure), for dissolution of crystalline phase of Mullite and Quartz is the need of the time to avoid the presence of residual ash in the final product and maximize the synthesis yield of better grade and applications.

Based on the challenges discussed above, the main questions which should come in the researchers' mind are: what is the optimum model of dissolution of fly ash particles? and, how alkali solutions make it happen?

The answer to the first question comes from the sinking core model of dissolution, proposed by Brouwers and Vaneijk [48]. This model has been useful in understanding the effect of hydroxyl ion on the ash particles. In fact, without going into the complex mathematical formulations, it is prudent to give more importance to the concept of dissolution. With this in view, pH (>12) of the alkaline environment has been found to be the most suitable parameter for optimum dissolution at which OH^- ions diffuse faster towards the reacting surface from the outer boundary to the central core region of the particles. This can be attributed to the release of glass ions (viz., SiO_3^{2-} and AlO_2^-) into the surrounding liquid. Moreover, the porosity of the particle developed on the reacting surfaces, can be helpful for effective diffusion of alkaline solution into the inner region. Further, it is interesting to note that the reactivity of inner region is more than the outer region.

In line with this, there is another model (Lattice Boltzmann model) worth mentioning here. This model is based on the assumption that the dissolution has to be diffusion controlled which can be attributed to the higher rate of chemical reaction than that of the diffusion [49]. Another dissolution model, presented by Hsu et al. [50], has been found to be quite useful to describe three different steps associated with the dissolution of solute (i.e., fly ash) particles such as, diffusion of solid-solid interface to the solid-liquid interface, diffusion of solute molecules from the solid-liquid interface through the surface layer up to its outer boundary and the diffusion of solute molecules from the outer surface layer to the bulk liquid phase.

Keeping all these challenges in mind, efforts are required to devise modified hydrothermal and fusion methods, where NaOH solution should be used only once in the primary activation step. The filtrate solutions obtained from the hydrothermal activation should be reused for further activation of the activated fly ash (the residues).In this context, efforts have been made to portray mechanism of zeolitization of fly ash in Chap. 4. Furthermore, innovation aspect in the conventional mechanism of zeolitization of fly ash has been touched by resorting to a novel methodology; three step experimentation and investigation, which have been presented in Chap. 5.

3.3 Conclusions

The state-of-the-art for synthesis of fly ash zeolites from fly ash employs various chemicals like NaOH, KOH, Na_2CO_3, KNO_3, $NaNO_3$, NH_4F and NH_4NO_3 as well as different techniques like hydrothermal, fusion, combination of both fusion and hydrothermal, molten salt method, microwave irradiation method. However, these methods yield final end products blended with zeolites, which comprise of some unreacted fly ash residues as impurities in the synthesized zeolites. The researchers have discussed various types of such impurities which warrant further activation for better end products. In view of such scenario, the authors have attempted to critically evaluate the state-of-the-art techniques and focus for further reactivation of the impurities in fly ash zeolites for their value addition. This endeavor is in favor of innovations in the conventional techniques and devises a novel methodology for improved fly ash zeolites, as discussed in Chap. 5.

References

1. Rayalu, S., Meshram, S.U., Hasan, M.Z.: Highly crystalline faujasitic zeolites from fly ash. J. Hazard. Mater. **B77**, 123–131 (2000)
2. Kumar, P., Mal, N., Oumi, Y., Yamanaa, K., Sanoc, T.: Mesoporous materials prepared using coal fly ash as the silicon and aluminium source. J. Mater. Chem. **11**, 3285–3290 (2001)
3. Murayama, N., Yamamoto, H., Shibata, J.: Mechanism of zeolite synthesis from coal fly ash by alkali hydrothermal reaction. Inter. J. Miner. Process. **64**, 1–17 (2002)
4. Kolay, P.K., Singh, D.N.: Characterization of alkali activated lagoon ash and its application for heavy metal retention. Fuel **8**, 483–489 (2002)
5. Adamczyk, Z., Bialecka, B.: Hydrothermal synthesis of zeolites from polish coal fly ash. Pol. J. Environ. Stud. **14**(6), 713–719 (2005)
6. Inada, M., Eguchi, Y., Enomoto, N., Hojo, J.: Synthesis of zeolite from coal fly ashes with different silica-alumina composition. Fuel **84**, 299–304 (2005)
7. Tanaka, H., Eguchi, H., Fujimoto, S., Hino, R.: Two-step process for synthesis of a single phase Na-A zeolite from coal fly ash by dialysis. Fuel **85**, 1329–1334 (2006)
8. Fukui, K., Katoh, M., Yamamoto, T., Yoshida, H.: Utilization of NaCl for phillipsite synthesis from fly ash by hydrothermal treatment with microwave heating. Adv. Powder Technol. **20**, 35–40 (2009)
9. Shigemoto, N., Hayashi, H.: Selective formation of Na-X zeolite from coal fly ash by fusion with sodium hydroxide prior to hydrothermal reaction. J. Mater. Sci. **28**, 4781–4786 (1993)
10. Berkgaut, V., Singer, A.: High capacity cation exchanger by hydrothermal zeolitization of coal fly ash. App. Clay. Sci. **10**, 369–378 (1996)
11. Scott, M.A., Kathleen, A.C., Dutta, P.K. Handbook of zeolite science and technology. CRC Press, NewYork. ISBN: 0-8247-4020-3 (2003)
12. Rayalu, S.S., Udhoji, J.S., Meshram, S.U., Naidu, R.R., Devotta, S.: Estimation of crystallinity in fly ash-based zeolite-A using XRD and IR spectroscopy. Curr. Sci. **89**(12), 2147–2151 (2005)
13. Jala, S., Goyal, D.: Fly ash as a soil ameliorant for improving crop production- a review. Bioresource. Technol. **96**, 1136–1147 (2006)
14. Somerset, V., Petrik, L., Iwuoha, E.: Alkaline hydrothermal conversion of fly ash precipitates into zeolites: the removal of mercury and lead ions from wastewater. J. Environ. Manage. **87**(1), 125–131 (2008)
15. Park, M., Choi, C.L., Lim, W.T., Kim, M.C., Choi, J., Heo, N.H.: Molten-salt method for the synthesis of zeolitic materials I; Zeolite formation in alkaline molten-salt system. Micropor. Mesopor. Mater. **37**, 81–89 (2000)
16. Park, M., Choi, C.L., Lim, W.T., Kim, M.C., Choi, J., Heo, N.H.: Molten salt method for the synthesis of zeolitic materials, II. Characterization of zeolitic materials. Micropor. Mesopor. Mater. **37**, 91–98 (2000)

17. Nugteren, H.W., Moreno, N., Sebastia, E., Querol, X.: Determination of the available Si and Al from coal fly ashes under alkaline conditions with the aim of synthesizing zeolites products. In: International Ash Utilization Symposium. Centre for Applied Energy Research, University of Kentuchy, Paper No. 71 (2001)
18. http://www.uio.no/15kjm5100_2008/solvothermal_a.pdf. Accessed on 20 June 2010
19. Kim, W., Seung, H., Ahn, B.J.: Synthesis of Na-P1 zeolite from coal fly ash. J. Indus. Eng. Chem. **3**(3), 185–190 (1997)
20. Querol, X., Morenom, N., Umanm, J.C., Juan, R., Hernandez, S., Fernandez, P.C., Ayora, C., Janssen, M., Garcıa, J.M.J., Linares, S.A., Cazorla, A.D.: Application of zeolitic material synthesized from fly ash to the decontamination of waste water and flue gas. J. Chem. Technol. Biotechnol. **77**, 292–298 (2002)
21. Querol, X., Moreno, N., Uman, J.C., Alastuey, A., Hernandez, E., Lopez-Soler, A., Plana, F.: Synthesis of zeolites from coal fly ash: an overview. Inter. J. Coal. Geol. **50**, 413–423 (2002)
22. Querol, X., Alastuey, A., Moreno, N., Alvarez, A.E., Garcı´a-Sa´nchez, A., Cam, J., Ayora, C., Simon, M.: Immobilization of heavy metals in polluted soils by the addition of zeolitic material synthesized from coal fly ash. Chemosphere. **62**, 171–180 (2006)
23. Chung, S.H., Kim, S.S., Nam, Y.M., Kim, S.M., Lee, B.J.: Synthesis and application of fly ash derived zeolites for pyrolysis of polyolefins. J. Indus. Eng. Chem. **9**(2), 181–187 (2003)
24. Fukui, K., Nishimoto, T., Takiguchi, M., Yoshida, H.: Effects of NaOH concentration on zeolite synthesis from fly ash with a hydrothermal treatment method. KONA. **24**, 183–191 (2006)
25. Kim, J.K., Lee, H.D.: Effects of step change of heating source on synthesis of zeolite 4A from coal fly ash. J. Indus. Eng. Chem. **15**, 736–742 (2009)
26. Jha, B., Singh, D.N.: A review on synthesis, characterization and industrial application of fly ash zeolites. J. Mater. Edu. **33**(1–2), 65–132 (2011)
27. Jha, B., Singh, D.N.: Zeolitization characteristics of a fly ash from wet- and dry- disposal systems. Acta. Geotech. Slov. **2**, 63–71 (2012)
28. Jha, B., Padmakumar, G.P., Singh, D.N., Iyer, K.: Synthesis of zeolites by fly ash alkali interaction. In: Proceedings of Indian Geotechnical Conference, Cochi, India, 1089–1092 (2011)
29. Hollman, G.G., Steenbruggen, G., Janssen, J.M.: A two-step process for the synthesis of zeolites from coal fly ash. Fuel **78**, 1225–1230 (1999)
30. Kolay, P.K., Singh, D.N.: Physical, chemical, mineralogical and thermal properties of cenospheres from a ash lagoon. Cem. Conc. Res. **31**(4), 539–542 (2001)
31. Fernandez, P.C., Galiano, Y.L., Rodrıguez, P.M.A., Vale, J., Querol, X.: Utilisation of zeolitised coal fly ash as immobilizing agent of a metallurgical waste. J. Chem. Technol. Biotechnol. **77**, 305–310 (2002)
32. Singh, D.N., Kolay, P.K.: Simulation of ash water interaction and its influence on ash characteristics. Prog. Ener. Combus. Sci. **28**, 267–299 (2002)
33. Fukui, K., Nishimoto, T., Takiguchi, M., Yoshida, H.: Effects of NaOH concentration on zeolite synthesis from fly ash with a hydrothermal treatment method. Soc. Powder. Technol. **40**, 497–504 (2003)
34. Wu, D., Zhang, B., Yan, L., Kong, H., Wang, X.: Effect of some additives on synthesis of zeolite from coal fly ash. Inter. J. Miner. Process. **80**(2–4), 266–272 (2006)
35. Criado, M., Fernández-Jiménez, A., de la Torre, A.G., Aranda, M.A.G., Palomo, A.: An XRD study of the effect of the SiO_2/Na_2O ratio on the alkali activation of fly ash. Cem. Conc. Res. **37**, 671–679 (2007)
36. Derkowski, A., Franus, W., Waniak-Nowicka, H., Czimerova, A.: Textural properties vs. *CEC* and EGME retention of Na-X zeolite prepared from fly ash at room temperature. Inter. J. Miner. Process. **82**, 57–68 (2006)
37. Fansuri, H., Pritchard, D., Zhang, D.: Manufacture of low-grade zeolites from fly ash for fertilizer applications. QCAT Technology Transfer Centre, Technology Court, Pullenvale Qld 4069, Australia (2008)

38. Sulaiman, S.K.: Preparation of synthetic zeolites from coal fly ash. Sol. Stat. Sci. Technol. **16**(1), 107–113 (2008)
39. Wałek, T.T., Saito, F., Zhang, Q.: The effect of low solid/liquid ratio on hydrothermal synthesis of zeolites from fly ash. Fuel **87**, 194–199 (2008)
40. Czurda, K.A., Haus, R.: Reactive barriers with fly ash zeolites for in situ ground water remediation. App. Clay. Sci. **21**, 13–20 (2002)
41. Inada, M., Tsujimoto, H., Eguchi, Y., Enomoto, N., Hojo, J.: Microwave-assisted zeolite synthesis from coal fly ash in hydrothermal Process. Fuel **84**, 1482–1486 (2005)
42. Tanaka, H., Fujii, A., Fujimoto, S., Tanaka, Y.: Microwave-assisted two-step process for the synthesis of a single-phase Na-A zeolite from coal fly ash. Adv. Powder Technol. **19**, 83–94 (2008)
43. Ojha, K., Pradhan, N.C., Samanta, A.N.: Zeolite from fly ash synthesis and characterization. Bull. Mater. Sci. **27**(6), 555–564 (2004)
44. Rungsuk, D., Apiratikul, R., Pavarajarn, V., Pavasant, P.: Zeolite synthesis from coal-fired power plant by fusion method. In: The 2nd International conference on sustainable energy and environment, 21–23 Nov, Bangkok, Thailand (2006)
45. Grutzeck, M.W., Siemer, D.D.: Zeolites synthesized from class F fly ash and sodium aluminate slurry. J. Am. Ceram. Soc. **80**(9), 2449–2453 (1997)
46. Molina, A., Poole, C.: A comparative study using two methods to produce zeolites from fly ash. Miner. Eng. **17**, 167–173 (2004)
47. Yao, Z.T., Xia, M.S., Ye, Y., Zhang, L.: Synthesis of zeolite Li-ABW from fly ash by fusion method. J. Hazard. Mater. **170**, 639–644 (2009)
48. Brouwers, H.J.H., Vaneijk, R.J.: Fly ash reactivity: Extension and application of a shrinking core model and thermodynamic approach. J. Mater. Sci. **37**, 2129–2141 (2002)
49. Shan, X., Doolen, G.: Multi component lattice-Boltzmann model with inter-particle interaction. J. Stat. Phy. **81**(1–2), 379–393 (1995)
50. Hsu, W.L., Lin, M.J., Hsu, J.P.: Dissolution of solid particle in liquids: A shrinking core model. Inter. J. Chem. Bio. Eng. **2**(4), 205–210 (2009)

Chapter 4
Mechanism of Zeolitization of Fly Ash

Abstract Fly ash is a matrix of several metal oxides, which have different molecular and structural properties and hence its interaction with NaOH is a complex (chemical) phenomenon. As such, synthesis of the fly ash zeolites, and their characteristics, is expected to depend on various attributes (viz., physical, chemical, mineralogical and morphological) of the fly ash. In order to realize the mechanism of the fly ash zeolitization, it would be quite prudent to picturize the fly ash particles and investigate its interaction with alkali, and interrelate the alkali activated fly ash with zeolites in terms of their mineralogical composition. Apart from this, the mechanism of formation of sodium aluminosilicates (the so called fly ash zeolites), after the interaction of the NaOH on the surface and the inner core of the fly ash particle, has been explained in the following.

Keywords Ash-alkali interaction · Dissolution · Nucleation · Gelation · Crystallization · Zeolitization

4.1 Modelling of the Fly Ash Particle

In general, majority of the fly ash particles have been reported to be spheroids in shape and efforts have been made to establish their chemical and mineralogical composition [1–5]. The outer most surface of the ash particle is different than the inner core, in terms of its reactivity with NaOH. In order to know more about the intra-particle heterogeneity of a fly ash particle, it can be picturized as depicted in Fig. 4.1.

It can be noticed that the smooth outer boundary (i.e., reacting surface) of the spherical particle (i.e., cenospheres) is mostly composed of the glassy material, which has been reported to exhibit comparatively higher reactivity than the Quartz and Mullite, embedded in the inner portion of the fly ash particle [5].

B. Jha and D.N. Singh, *Fly Ash Zeolites*, Advanced Structured Materials 78,
DOI 10.1007/978-981-10-1404-8_4

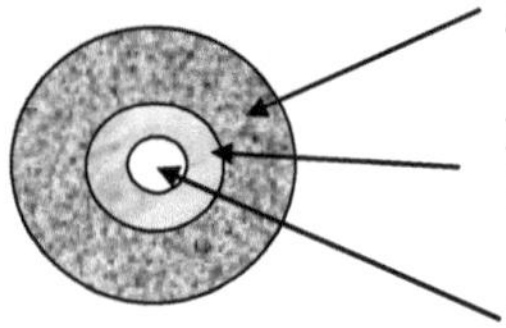

Fig. 4.1 The cross-section and composition of a micro spherical fly ash particle

4.2 Chemical Reaction Potential of the Fly Ash

The outermost layer of the fly ash could be considered to be made up of an easily degradable ingredient (the glass) under alkaline condition. Based on this fact, it has been demonstrated that the activation of glassy ingredients contributes much to the dissolution stage reactions. The effect of alkali on the amorphous alumino-silicate content in glassy ingredients of the fly ash is tremendously intense. This can be attributed, mainly, to the presence of amorphous silica and aluminium in it.

In addition, the amorphous silica and alumina, from the reacting surface of the glass layer, get dissolved in the alkaline solution thereby paving way for the diffusion of alkaline solution to the inner layers of the particles. In fact, the crystalline phases of SiO_2 present in the Quartz are attacked by cation (i.e., Na^+) as a mineralizer to get transformed into the soluble form of the sodium silicate, which results in a significant increase in Si^{4+} in the solvent. As the alkaline solvent penetrates the inner core of the fly ash particle, the most stable crystalline forms of SiO_2 and Al_2O_3 from Mullite also get dissolved, which finally results in comparatively an increased concentration of Al^{3+} than Si^{4+} in the solvent solution. This creates a favorable condition for the zeolitization of the fly ash and the alkaline reactions undergoing on the reacting surfaces of the fly ash particles, during their dissolution, contribute a lot to the process of gelation, as described by Eqs. 4.1 and 4.2 [6].

$$Si(OH)_4 + H_2O + OH^- \Leftrightarrow (H_2O)\ Si(OH)_5^- \tag{4.1}$$

$$2Al(OH)_4^- \Leftrightarrow \left[(OH)_3Al{-}O{-}Al(OH)_3\right]^{2-} + H_2O \tag{4.2}$$

In order to investigate the mechanism of zeolitization of the fly ash by alkali (NaOH or KOH), it is of utmost importance to study the characteristics of the raw fly ash (viz., type of the fly ash sample, its chemical and mineralogical composition, and cross section of its particles) [6–8]. With this in view, Quartz (SiO_2) and Mullite ($Al_6Si_2O_{13}$) have been considered as the main crystalline ingredients of the fly ash. Although, their total quantity is less in the fly ash, a wide range of the variation can be observed corresponding to the type of the samples presented in Table 4.1. It can be noticed that the majority of the fly ash samples contains amorphous glass fraction which consists of amorphous forms of SiO_2 and Al_2O_3

Table 4.1 Variation in the major mineralogical composition of the coal fly ash

Quantity of minerals (wt%)			References
Quartz	Mullite	Glass	
–	–	60–80	[10, 11]
2.2	27.2	43.7	[12]
17.6	41.2	33	
6.5	13.7	76	[13]
3.4	40.4	51.8	[4]
8.6	19.4	62.7	[14]
8.6	19.4	62.7	[15]
6.6	3.8	85.6	
4	15	70	[16]

– Not available

associated with the glass, which in turn is composed of several oxides (viz., SiO_2; network formers Al_2O_3, Fe_2O_3, TiO_2, P_2O_5 and the network modifiers CaO, MgO, Na_2O and K_2O) [9]. With this in view, it has been opined that a wide variation in the mineralogical composition of the fly ash (refer Table 4.1) can be correlated with its chemical composition and hence its Si/Al ratio (refer Fig. 4.2a).

In order to check the suitability of the fly ash for its zeolitization, data available in literature [3], for 23 fly ash samples, procured from Spain, Netherlands, Italy and Greece, with a wide variation in their chemical (i.e., with reference to SiO_2 and AlO_2 contents) and mineralogical compositions (i.e., with reference to Quartz, Mullite and glass content), was synthesized, with an intention to develop a correlation between Si/Al ratio and the amount of silica extracted during dissolution process (in g/kg). A random variation between these parameters was observed as depicted in Fig. 4.2a [3]. It can be observed that Si/Al ratio of 2.8 corresponds to the ratio of glass to total quantity of Quartz and Mullite equal to 2.36 (as highlighted by dotted lines). However, as depicted in Fig. 4.2b, corresponding to Si/Al ratio of 2.8, the amount of

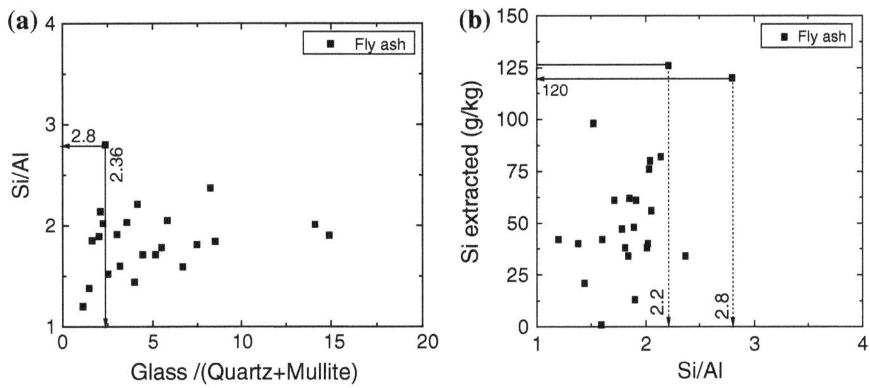

Fig. 4.2 Variation of **a** Si/Al ratio of fly ash samples with glass/(Quartz + Mullite) ratio and **b** the amount of silica extracted due to alkali activation with Si/Al ratio

Si extracted is 120 g/kg i.e., slightly lesser than the maximum value (126 g/kg). This indicates that the fly ash sample with Si/Al ratio equal to 2.8 can be graded as an optimum sample from the point of view of synthesis of ash zeolites by its alkali activation, where the amount of extracted Si plays an important role in nucleation and crystallization of zeolite crystals. Furthermore, the amount of silica extracted from the fly ash into the alkali solvent will depend upon the chemical composition of particles and their reactivity with the alkali.

The variations in products of the above reactions can be controlled by opting desired variations in the molarity of alkali (e.g., NaOH), SiO_2/Al_2O_3 ratio of the fly ash, reaction time and temperature. It has been demonstrated that an increase in *pH* (i.e., >10) of the solution can increase the solubility of the amorphous silica in the alkaline solution. This can contribute to an increase in dissolution of more monomeric silica and aluminum complexes leading to the formation of negatively charged ion complexes as presented by the product side of Eqs. 4.1 and 4.2.

Further, with reference to above reactions, zeolitization of the fly ash can be attributed to an equilibrium stage of hydroxylation reactions between the solution and the fly ash. In fact, by providing proper conditions for reactions (viz., temperature, concentration and time), the solubility of silica and aluminum ions can be increased so that their condensation reactions between pairs of different ions (viz., SiO_3^{2-}, AlO_2^-, Na^+, Fe^{3+} and $HTiO_3^-$) results in the formation of crystal nuclei. The further growth of crystals can be attributed to the association of all available nuclei, lying either around the activated fly ash residue particles or in the solution, leading to crystallization of different zeolite phases. It can be noted that zeolitization of the fly ash results in replacement of Si^{4+} by Al^{3+} in some of the silica tetrahedra of sodium silicate framework, which can be attributed to development of negative charge on the surface of the framework, which in turn can be neutralized by the attachment of either any one or the group of cations (viz., Na^+, K^+, Ca^{2+}, Fe^{3+}) from the solvent solution.

In fact, replacement of silica by aluminum in the tetrahedron can cause differentiation in the redistribution of the electric charge between the Al-O and Si-O bonds, which result in polarization of the chemical bonds and the enhancement of the chemically active centers (of positive and negative charge) in the crystal lattice of ash zeolites. This can be attributed to the development of terminal groups (viz., ≡Si-OH, ≡Si-O-Na, ≡Si-O-, $(\equiv Si\text{-}O)_3$ and Al-O-) in the alkaline solvent, which can lead to the synthesis of more complex products (i.e. zeolites).

As discussed above, the formation of zeolites can mainly be attributed to the liberation of aluminum from the fly ash (≈40 % by Mullite and 60 % by glass) to the liquid phase during its alkaline activation. In addition, the presence of less reactive ingredients viz., Hematite (i.e., Fe_2O_3, which is about 0.3–1.5 % by weight in the fly ash), Magnetite (i.e., Fe_3O_4, which is about 0.5–2.5 % by weight of the fly ash) and some trace elements (viz., As, Sr, Cd, Zn, Pb, Cs, Ti, Mn, Ca, Mg etc.) in the glassy fraction of the particles can get accommodated in the zeolites framework, as impurity. However, degree of purity of the ash zeolites can vary significantly with the change in methods of synthesis and the type of raw fly ash employed for the purpose [3, 4, 13, 17, 18].

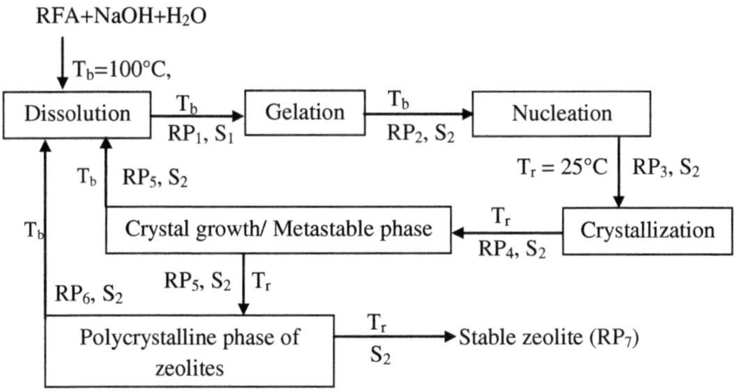

Fig. 4.3 Flowchart for various stages in zeolitization of fly ash by hydrothermal method

Figure 4.3 presents the basic concept of the mechanism of the zeolite formation. This figure summarizes the complete process of zeolitization and the sequence of various types of chemical transformations, the input and the output of each of the intermediate processes and the formation of their products.

It can be noticed from Fig. 4.3 that a variety of continuous processes are involved in zeolitization of the raw fly ash (RFA). The initial process of dissolution is reported to have hydrothermal temperature of boiling water (i.e., T_b = 100 °C) at atmospheric pressure in open system under a specific *pH* of the solution. The output of this stage is the activated solid phase (i.e., the product, designated by RP_1) with less quantity of silica and alumina and some new minerals. In fact, the loss of silica and aluminium from the fly ash can contribute to the increase of these ions in the solution (i.e., S_1). As a result of this, the precipitation reactions between available ions in solution S_1 can initiate the formation of gel of the reaction products i.e., solid phase RP_2 (negatively charged ion complexes as depicted in Eqs. 4.1 and 4.2) and resulting solution S_2. It is to be noted here that the reduction in the concentration of Al^{3+} (i.e., comparatively more) and Si^{4+} in the solution S_2, as compared to their initial quantity available in the soluble form in solution S_1, is mostly associated with this stage. The input for the nucleation stage consists of a mixture of the solution S_2 and solid residue RP_2, maintained at the same temperature T_b, which can contribute to the growth of the crystal nuclei in its output stream (i.e., solid phase RP_3). Whereas, the solution phase S_2 can maintain the required alkalinity in the process without undergoing significant changes in its chemical composition.

In order to allow the system to mature at room temperature (designated by T_r), a change in process such as inclusion of a relaxation time i.e., ageing period, can initiate and further expedite the agglomeration of the crystal nuclei during crystallization stage, which can favour the crystal growth (i.e., solid phase RP_4) in the liquid phase S_2). The output of this stage can consist of many crystals still in the growing stage with less stability of their bonds (viz., Si-O-Al and Si-O-Si) i.e. referred to as their metastable stage. As the ageing duration is extended further, the prolonged process of crystallization can result in the further growth in the metastable crystals (RP_5). This can be attributed to the birth of new crystals of varying dimensions and shapes (i.e., known as polycrystalline phase) which after prolonged ageing and further reactions, can grow as more stable crystals of zeolites which can be finally separated from the solution (S_2.). Furthermore, it is worth mentioning that the output of each process can contain impurities in the form of trace elements and/or unreacted residue of fly ash particles which can be attributed to the incomplete reaction in that process. With this in view, the reactivation or recycling of the outputs (i.e., solid residues R_5 and R_6) with the resulting solutions S_2 as supernatants can result in lowering or removal of previously incorporated impurities from the final output and synthesis of more stable and purer form of fly ash zeolites of a specific morphology.

For example, to understand the mechanism of development of morphology of the synthesis product in the hydrothermal method, the type of alkali source employed and its reactivity with the fly ash, can influence the input and output streams of each as depicted in Fig. 4.4. In fact, alkali activation of the fly ash particles, with NaOH, causes etching of the outer surface of its particles and increases its surface roughness because of the large scale dissolution of Si^{4+} and Al^{3+} in the alkali solution. This can also be attributed to the precipitation reaction products, their nucleation and subsequent crystallization, as fine crystals of zeolite P, mostly seen as surface deposits in the activated fly ash residue particles (refer Fig. 4.4a) [12]. The increase in the rate of dissolution can be directly correlated with the increase in alkali concentration and/or temperature, which can finally result in the rapid nucleation and crystallization of big spherical crystals of Sodalite, which has been shown as projecting out of the surface as seen in Fig. 4.4a.

Further, alkali activation of the fly ash with Na_2CO_3 can result in a coating of the thin film of zeolite P around the fly ash particles (refer to Fig. 4.4b), whereas, that with KOH can develop a surface deposit of egg shaped, ellipsoidal crystals of Chabazite (refer to Fig. 4.4c). Based on the variation in the developed morphologies, it can be confirmed that different types of alkalis can display their superiority with reference to

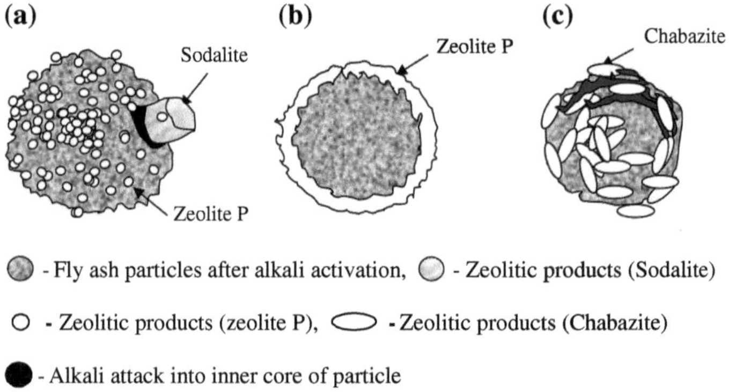

Fig. 4.4 Pictorial representation of the cross section of the fly ash particle undergoing alkali activation with **a** NaOH, **b** Na_2CO_3 and **c** KOH and development of the surface morphology by hydrothermal method

their degree of zeolitization and type of the crystal morphology synthesized, which can be assigned an increasing order as NaOH > KOH > Na_2CO_3 [12].

Further, it has been opined that the microwave heating in the hydrothermal method is effective in exciting the particles at molecular level, fast dissolution of silica and aluminium ions in the solution and reducing the total synthesis time as depicted in Table 4.1. Similarly, the fusion and hydrothermal methods of synthesis have an additional benefit of evolving a new process of thermal treatment (i.e., dry mixing of the fly ash and alkali, in the initial stage of activation). In fact, the high temperature heating of alkali, can facilitate its fusion with the fly ash, which can ultimately result in the extraction of silica and aluminium in dry stage. Furthermore, the dry stage reaction of extracted silica and aluminium with the Na^+ can result in precipitation of sodium silicates and sodium aluminates in the product of fusion. In continuation, the second stage of synthesis is in accordance with the hydrothermal method of alkali activation in wet stage, as discussed above. In fact, alkali activation of the fly ash particles results in dissolution of silica and alumina into the solution as discussed above. However, it is interesting to find a wide variation in concentration of silica and aluminium ions in the solution with the activation time and simultaneous crystallization of different zeolites, corresponding to different time, as depicted in Fig. 4.5[19].

It can be noticed from Fig. 4.5 that the variation of concentration of soluble silica in the solution sharply increases up to 10,000 ppm with only 6 h of alkali activation and thereafter it decreases up to 6400 ppm, corresponding to 20 h of treatment. Further, an equilibrium concentration of silica in the solution is achieved corresponding to wide variations in activation time (from 20 to 96 h). Moreover, the trend of variation in the concentration of aluminium ion in the solution is found to increase initially up to 460 ppm, corresponding to 4 h of activation, which later decreases sharply up to 20 ppm, corresponding to 12 h of activation and finally gets stabilized and equilibrated. It is interesting to observe that there is no increase in the concentration of silica and aluminium ions in the solution despite continuous activation of the fly ash particles (for 20 to 96 h). This can be attributed to

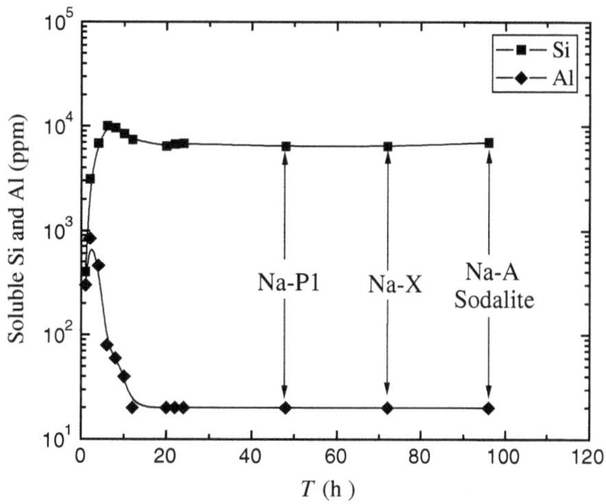

Fig. 4.5 Dissolution of fly ash and crystallization of zeolites in hydrothermal method

incorporation of all freshly dissolved quantities of these ions in the precipitation products, corresponding to alkali activation for the said duration. Also, depending upon the Si/Al ratio, corresponding to their excess (i.e., over equilibrium concentration as depicted in Fig. 4.5) concentration in the solution, different types of zeolites (viz., Na-P1, Na-X and Na-A), can get crystallized, corresponding to different reaction times (viz., 48, 72 and 96 h, respectively) [20]. However, the amount of yield of conversion of fly ash to zeolites has been reported to vary from 40–60 %, which can be attributed to the following limitations [12, 15].

1. Incomplete conversion—as the crystallization process is faster than the dissolution process, all the leached out Al in the solution can get associated in the formation of gel precursor before crystallization starts over the surface of the fly ash particles leading to the reduction in the available surface for further dissolution. Finally, the remaining ash surface is getting blocked by the precipitates of ash-alkali interaction and hence the dissolution process gets hindered [3, 5, 19, 21].
2. The presence of non-reactive components such as iron oxides, Calcite, Lime, Gypsum, Mullite, non-reactive glass and trace elements (e.g., Hg, Cu, Pb, Ni, Cd, Se, B, As, Zn, Ba, Sr, Fe, Mn, Mg, Rb, Mo, Sb, Cs, Li, Ce, Nd, Sb) in the ash, which can increase the quantity of impurities in the solution and in turn delay the crystallization of zeolites in the final products [3–5, 13, 19, 22].
3. The Si/Al ratio of the solution can only be influenced by the synthesis conditions (viz., temperature, liquid to solid (L/S) ratio, molarity of the alkali, types of the alkali and fly ash) employed, which can significantly affect the chemical and mineralogical compositions of the synthesized zeolites [3, 5, 13, 22–24]. A summary of the detailed study of the available literature is presented in Table 4.2 to exhibit the effect of variation in L/S ratio, temperature, activation time and molarity of the alkali solution on the final yield (i.e., in terms of quantity of the product as weight % of fly ash) of zeolitization of the fly ash.

Table 4.2 Effects of optimized L/S ratio, activation time (*T*), type of alkali solution and Temperature on hydrothermal conversion of fly ash to zeolites

L/S (ml/g)	*T*(h)	*M*(CT)[a]	Temp. (°C)	Yield[b] (%)	Reference
10	48–96	4–5 (Na^+)	90	50	[11]
4–7	6–48	1.16 (Na^+)	80–320	40	[25]
8	12–48	3.5 (Na^+)	100	40	[26, 27]
10	24–48	2–3 (Na^+)	100–150	40	[10]
3,20	6	2 (Na^+)	90, 50	–	[3]
6.7–20	4	–	104	–	[28]
25	12–25	1–2.5 (Na^+)	40–70	–	[29]
2	8–24	2–3	125–150	40–55	[15]
3.5	–	–	–	45	[16]
15	72	1(K^+)	160	–	[30]

[a]*M* molarity of solution, *CT* type of cation present in the solution
[b]Calculated on the basis of *CEC* of the synthesized fly ash zeolites in comparison to that of pure Na-P1 (≈500 meq/100 g)

4.3 Conclusions

The mechanism of zeolitization of the fly ash by resorting to alkali activation depends on various parameters (viz., molarity of alkali solution, time of activation, temperature, physical, chemical, mineralogical and morphological properties of the fly ash and its activated residues). The Si/Al ratio, corresponding to various mineral phases present in the fly ash, controls its reactivity with the alkalies. The dissolution of Si and Al of the fly ash in the solution facilitates interaction between these ions, which finally undergo nucleation to form new products of Si and Al soluble ions. Subsequently, gelation and crystallization of the newly formed products in the solution leads to zeolitization of the fly ash. Specially, dissolution of Al of the fly ash with reacting solution plays a decisive role in its zeolitization. The mechanism of zeolitization includes nucleation and gelation of both the soluble Si in the solution, followed by incorporation of Al in the gel and finally crystallization to form zeolites. The crystals of the fly ash zeolites get deposited on the unreacted ash particle, which is an outcome of inappropriate alkali activation of the minerals present in the fly ash.

References

1. Mortier, W.J.: Zeolite electronegativity related to physicochemical properties. J. Catalysis. **55**, 138–145 (1978)
2. Sand, L.B.: Zeolite synthesis and crystallization. Pure. App. Chem. **52**, 2105–2113 (1980)
3. Nugteren, H.W., Moreno, N., Sebastia, E., Querol, X.: Determination of the available Si and Al from coal fly ashes under alkaline conditions with the aim of synthesizing zeolites products. International Ash Utilization Symposium, Centre for Applied Energy Research, University of Kentuchy, Paper no. 71 (2001)
4. Fernandez, P.C., Galiano, Y.L., Rodrıguez, P.M.A., Vale, J., Querol, X.: Utilisation of zeolitised coal fly ash as immobilising agent of a metallurgical waste. J. Chem. Technol. Biotechnol. **77**, 305–310 (2002)
5. Inada, M., Eguchi, Y., Enomoto, N., Hojo, J.: Synthesis of zeolite from coal fly ashes with different silica-alumina composition. Fuel **84**, 299–304 (2005)
6. Inada, M., Tsujimoto, H., Eguchi, Y., Enomoto, N., Hojo, J.: Microwave-assisted zeolite synthesis from coal fly ash in hydrothermal Process. Fuel **84**, 1482–1486 (2005)
7. Fukui, K., Nishimoto, T., Takiguchi, M., Yoshida, H.: Effects of NaOH concentration on zeolite synthesis from fly ash with a hydrothermal treatment method. KONA. **24**, 183–191 (2006)
8. Tanaka, H., Fujii, A., Fujimoto, S., Tanaka, Y.: Microwave-assisted two-step process for the synthesis of a single-phase Na-A zeolite from coal fly ash. Adv. Powder Technol. **19**, 83–94 (2008)
9. Towler, M.R., Stanton, K.T., Mooney, P., Hill, R.G., Moreno, N., Querol, X.: Modelling of glass phase in fly ashes using network connectivity theory. J. Chem. Technol. Biotechnol. **77**, 240–245 (2002)
10. Kim, W., Seung, H., Ahn, B.J.: Synthesis of Na-P1 zeolite from coal fly ash. J. Indus. Eng. Chem. **3**(3), 185–190 (1997)
11. Scott, M.A., Kathleen, A.C., Dutta, P.K.: Handbook of Zeolite Science and Technology. CRC Press, NewYork (2003). ISBN:0-8247-4020-3

12. Murayama, N., Yamamoto, H., Shibata, J.: Mechanism of zeolite synthesis from coal fly ash by alkali hydrothermal reaction. Inter. J. Miner. Process. **64**, 1–17 (2002)
13. Shigemoto, N., Hayashi, H.: Selective formation of Na-X zeolite from coal fly ash by fusion with sodium hydroxide prior to hydrothermal reaction. J. Mater. Sci. **28**, 4781–4786 (1993)
14. Querol, X., Alastuey, A., Moreno, N., Alvarez, A.E., García-Sánchez, A., Cam, J., Ayora, C., Simon, M.: Immobilization of heavy metals in polluted soils by the addition of zeolitic material synthesized from coal fly ash. Chemosphere. **62**, 171–180 (2006)
15. Querol, X., Moreno, N., Alastuey, A., Juan, R., Andres, J.M., Lopez-Soler, A., Ayora, C., Medinaceli, A., Valero, A.: Synthesis of high ion exchange zeolites from coal fly ash. Geologica. Acta. **5**(1), 49–57 (2007)
16. Steenbruggen, G., Hollman, G.G.: The synthesis of zeolites from fly ash and the properties of the zeolite products. J. Geochem. Explor. **62**, 305–309 (1998)
17. Kumar, P., Mal, N., Oumi, Y., Yamanaa, K., Sanoc, T.: Mesoporous materials prepared using coal fly ash as the silicon and aluminium source. J. Mater. Chem. **11**, 3285–3290 (2001)
18. Derkowski, A., Franus, W., Waniak-Nowicka, H., Czimerova, A.: Textural properties vs. CEC and EGME retention of Na-X zeolite prepared from fly ash at room temperature. Inter. J. Miner. Process. **82**, 57–68 (2006)
19. Hollman, G.G., Steenbruggen, G., Janssen, J.M.: A two-step process for the synthesis of zeolites from coal fly ash. Fuel **78**, 1225–1230 (1999)
20. Criado, M., Fernández-Jiménez, A., de la Torre, A.G., Aranda, M.A.G., Palomo, A.: An XRD study of the effect of the SiO_2/Na_2O ratio on the alkali activation of fly ash. Cem. Conc. Res. **37**, 671–679 (2007)
21. Querol, X., Moreno, N., Uman, J.C., Juan, R., Hernandez, S., Fernandez, P.C., Ayora, C., Janssen, M., Garcıa, J.M.J., Linares, S.A., Cazorla, A.D.: Application of zeolitic material synthesised from fly ash to the decontamination of waste water and flue gas. J. Chem. Technol. Biotechnol. **77**, 292–298 (2002)
22. Somerset, V.S., Petrik, L.F., White, R., Klinka, M.J., Key, D., Iwuoha, E.I.: Alkaline hydrothermal zeolites synthesized from high SiO_2 and Al_2O_3 co-disposal fly ash filtrates. Fuel **84**, 2324–2329 (2005)
23. Rayalu, S., Meshram, S.U., Hasan, M.Z.: Highly crystalline faujasitic zeolites from fly ash. J. Hazard. Mater. **B77**, 123–131 (2000)
24. Rungsuk, D., Apiratikul, R., Pavarajarn, V.,Pavasant, P.: Zeolite synthesis from coal-fired power plant by fusion method. In: The 2nd International Conference on Sustainable Energy and Environment, Bangkok, Thailand, 21–23 November 2006
25. Adamczyk, Z., Bialecka, B.: Hydrothermal synthesis of zeolites from polish coal fly ash. Pol. J Environ. Stud. **14**(6), 713–719 (2005)
26. Kolay, P.K., Singh, D.N.: Physical, chemical, mineralogical and thermal properties of cenospheres from a ash lagoon. Cem. Conc. Res. **31**(4), 539–542 (2001)
27. Kolay, P.K., Singh, D.N., Murti, M.V.R.: Synthesis of zeolites from lagoon ash. Fuel **80**, 739–745 (2001)
28. Wałek, T.T., Saito, F., Zhang, Q.: The effect of low solid/liquid ratio on hydrothermal synthesis of zeolites from fly ash. Fuel **87**, 194–199 (2008)
29. Fukui, K., Katoh, M., Yamamoto, T., Yoshida, H.: Utilization of NaCl for phillipsite synthesis from fly ash by hydrothermal treatment with microwave heating. Adv. Powder Technol. **20**, 35–40 (2009)
30. Qafoku, N.P., Ainsworth, C.C., Szecsody, J.E., Qafoku, O.S.: Aluminium effect on dissolution and precipitation under hyper alkaline conditions. J. Environ. Qual. **32**(6), 2354–2363 (2003)

Chapter 5
Novel Techniques for Synthesis and Characterization of Fly Ash Zeolites

Abstract As discussed in the previous chapters, conventional techniques of synthesis of fly ash zeolites have not been found successful in synthesizing zeolites of higher grades (i.e., zeolites possessing high cation exchange capacity), mainly due to incomplete zeolitization of the fly ash, the complexities associated with the liquid by-product and the impurities present in the activated residues. The degree of activation of the type fly ash may also depend on its zeolitization potential and hence identification of the most suitable fly ash which would yield products of improved grade becomes a prime focus. In this context, out of the two types of the fly ashes (viz., hopper ash and lagoon ash, available as dry powder at the electrostatic precipitator and as wet powder at lagoons, respectively), which have different characteristics (viz., physical, chemical, mineralogical and morphological), depending upon their disposal site conditions (dry and wet) were used in this study. Based on detailed experimentation, hopper ash has been ascertained to have faster reaction with NaOH and thus yields superior residue with higher cation exchange capacity than the lagoon ash. This could be observed from the X-ray fluorescence results, X-ray diffractograms and micrographs of the two ashes and their products, after hydrothermal treatment. Finally, it has been demonstrated that the hopper ash exhibits better zeolitization potential than the lagoon ash. Furthermore, to synthesize higher grade fly ash zeolites from the hopper ash, a technique which involves a very innovative synthesis process, has been developed and its details are presented in this chapter. Contrary to the conventional hydrothermal technique, this novel technique is based on 'three-step activation' of the hopper fly ash by employing hydrothermal as well as fusion activation and hence results in synthesis of zeolites of very high cation-exchange capacity.

Keywords Fly ash · Hopper ash · Lagoon ash · Alkali activation · Hydrothermal · Fusion · Grinding · Residues · Supernatant · Characteristics · Zeolitization

B. Jha and D.N. Singh, *Fly Ash Zeolites*, Advanced Structured Materials 78,
DOI 10.1007/978-981-10-1404-8_5

5.1 Materials

Raw fly ash (RFA) samples, used in the present study, were collected, separately, from hoppers (i.e., a dry disposal system in the electrostatic precipitator) and lagoons (i.e., wet disposal system), of Bhusaval Thermal Power Station (BTPS), Bhusaval, Maharashtra, India. The reagent grade caustic soda (NaOH, in pellet form) supplied by Thomas Baker, Mumbai, was used in this study for chemical activation of the fly ash, as described in the following. The deionized water was used to prepare a stock solution of NaOH (molarity, M, ranging from 0.5 to 3.5, with an increment of 0.5 unit). Furthermore, detailed methodology followed to characterize the end products is also presented in this chapter.

5.2 Alkali Activation of the Fly Ash

In order to synthesize zeolites from fly ash by its activation with NaOH, attempts have been made to identify a suitable fly ash out of its two disposal sites (viz., dry site at the electrostatic precipitator and wet site at the lagoons in the thermal power plants) for conventional (i.e., one step) hydrothermal activation technique [1–10]. Subsequently, the fly ash ascertained to exhibit improved zeolitization potential has been prefered to undergo novel hydrothermal treatment processes (viz., three step activation by hydrothermal technique and three step fusions) to activate the fly ash significantly for synthesis of fly ash zeolites with high cation exchange capacity [11–15]. The details of both the types of alkali activations (viz., conventional with the two ashes and three step activations with the superior ash) are presented in the following.

5.2.1 A Novel Hydrothermal Technique

The sample, in slurry form, was prepared by mixing a solution of NaOH and the RFA in a fixed proportion. To ensure significant zeolitization by hydrothermal technique, the slurry for all the treatments was maintained with a fixed liquid to solid ratio, $L/S \approx 10$ in ml/g [1–4]. In fact, the L/S as stated above has been reported to yield maximum conversion of RFA to FAZ up to 80 % [4].

In this context, the role of NaOH has been established by several researchers [2, 5–7]. Accordingly, the stock solution of NaOH, based on its high corrosive property, has been assumed to cause significant etching of the fly ash ingredients at the hydrothermal temperature of 100 °C. The experiment was carried out in a fabricated set up, which comprises of a 10 L reactor fitted with an open reflux condenser, thermostatically controlled heater and a magnetic stirrer. A schematic sketch of the experimental set up is presented in Fig. 5.1.

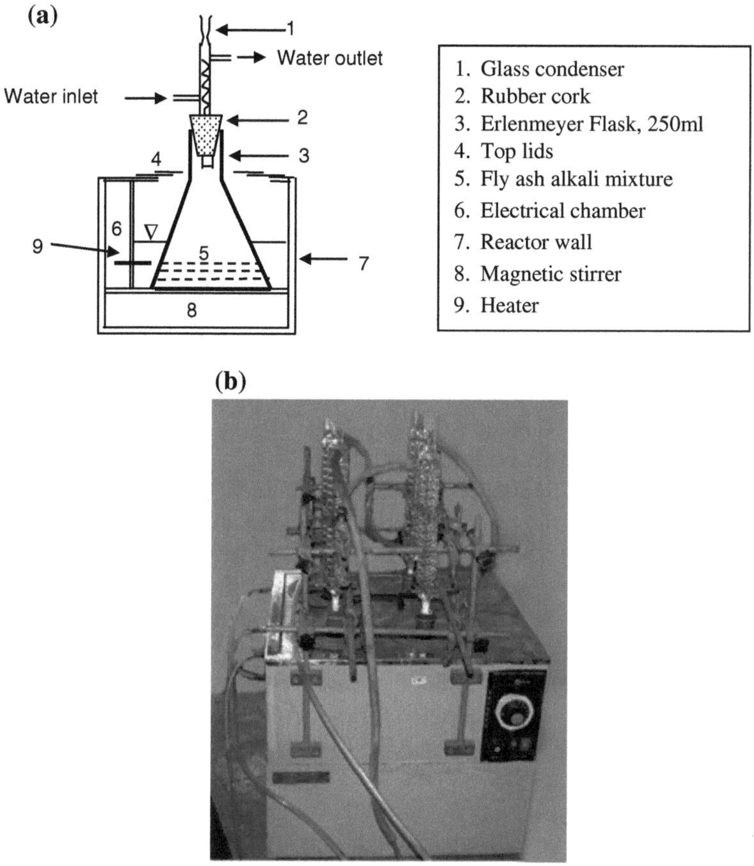

Fig. 5.1 Details of the hydrothermal reflux set up. **a** Schematic diagram and **b** Full set up in running condition

With an intention to solve the problem of separation of a FAZ from the bulk of the alkali activated fly ash (AAF) and residual ash (i.e. the impurity, designated by IMP) in primary treatment [8–10], modification in the zeolite synthesis process was carried out and its details are presented in this chapter. As reported by Jha and Singh [8], the yield of a pure and stable form of FAZ can be increased by complete digestion of IMP, present in the residue [5]. With this in view, the present study is based on the hypothesis that the alkali activation of the fly ash (i.e., SiO_2, Al_2O_3 in the form of glass, Quartz and Mullite) with NaOH solution, during each step of the treatment, partially dissolves many of these ingredients to form zeolites. However, the used solution (i.e., supernatant, designated by ST) is left over with considerable alkalinity at the end of the treatment.

In fact, it has been assumed that the repeated treatment of the residual ash by the ST should result in its further dissolution, decrease in IMP and increase in both quantity and purity of FAZ in the final AAF. In addition, it has been assumed that the repeated activation of the AAF should result in the formation of new zeolites, further growth of previously crystallized zeolites and transition of metastable zeolites to stable form [11–15].

In view of the above, this technique focuses to recycle/resuse initial products (i.e., the residue, AAF and the supernatant, ST) up to three steps of treatment, which has been termed as three step activation (TSA) by hydrothermal technique [12–15]. Moreover, uniform conditions of various parameters (viz., L/S, duration of treatment, temperature and pressure) were maintained during each step of the treatment. As such, with an intention to investigate the relative variation in the characteristics (viz., chemical, mineralogical and morphological) of the products of several recyclings, TSA techniques for advanced characterization of the end products have also been devised. This is a major modification in the process of zeolitization of fly ash, by the hydrothermal technique in this study, over conventional techniques demonstrated by the earlier researchers [2, 16].

In this context, to differentiate each step of treatment from others, Step-1 of the activation process (the primary treatment) has been designated as PT. Subsequently, first recycled treatment (i.e., Step-2 of the TSA) has been designated as R1, whereas, the final step i.e. second recycled treatment (i.e., Step-3 of the TSA) has been designated as R2. A typical flowchart for the step-wise synthesis of FAZ is presented in Fig. 5.2.

It can be noticed from Fig. 5.2 that there are series of processes involved in each step of the TSA. The processes adopted for the zeolite synthesis are: (a) preparation of the slurry (viz., fly ash-NaOH or AAF-ST), (b) the hydrothermal treatmentof the slurry, (c) aging of the activated slurry for 24 h, (d) separation of the residue (i.e., AAF) from STby centrifugation (at 2000 rpm), (e) washing the AAF with de-ionized water and (f) over-night oven drying at 100 ± 5 °C. In addition, based on the reported literature [17], it is learnt that a layer of zeolite deposited on the outer surface of the AAF of the Step-1, restricts further activation, dissolution and zeolitization of the residual ash. With this in view, the process of grinding the AAF, manually, by the use of mortar and pestle, is of much significance, before its further treatment, in the subsequent steps.

In fact, powdered AAF contains some FAZs' and IMP; hence there is enough scope for further activation of IMP and synthesis of zeolites by employing the TSA. To investigate the relative effect of recycling, on the quantity and purity of the products of the TSA, the hydrothermal treatment in all its steps (viz., PT, R1, and R2) was carried out for a constant duration (i.e., designated by T). In addition, to study the effect of the duration on the characteristics of the AAF, and the ST, the duration of the TSA was varied from 12 to 48 h (i.e., at an increment of 12 h).

The total durations allowed up to any of the three steps of the TSA can be quantified in terms of its cumulative duration (T_c). It is notable that, for each duration (T) of the TSA, T_c can be calculated as T, $2T$ and $3T$ corresponding to the steps PT, R1 and R2, respectively.

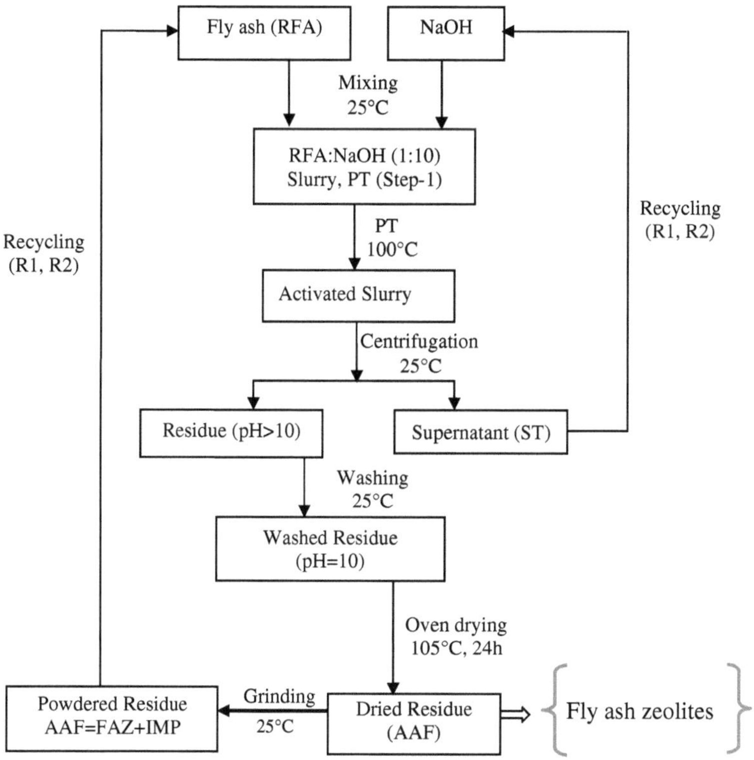

Fig. 5.2 The flowchart for three step activation of the fly ash by hydrothermal technique

Further, to investigate the effect of molarity of the stock solution of NaOH on the residue-supernatant (i.e., AAF-ST) interaction, variation of both *pH* and electrical conductivity (*EC*) of the ST was monitored throughout the TSA technique. A transition in both these parametersdue to each step of the TSA was given much significance, when the ST was employed as a reagent for R1 and R2 cycles. To ascertain the maximum yield of precipitation-cum-zeolitization reactions between the soluble ions (viz., Si^{4+}, Al^{3+}, Na^{+} and OH^{-}), the treated slurry was allowed to mature for an aging period of 24 h. In addition, ST of each step of the treatment was investigated for variation in its chemical composition (in terms of Si and Al contents). To simplify the interpretation of the results, the end products (viz., AAF and ST) of each step of the TSA, has been designated, as listed in Table 5.1.

As well, RFA was collected from lagoon of the thermal power plant. To differentiate the original hopper ash (OHA) and lagoon ash (OLA) from their residues obtained after alkali activation (viz., activated hopper ash, AHA, and activated lagoon ash, ALA) the samples have been designated as listed in Table 5.2.

Table 5.1 Sample designation for different types of hydrothermal activation of hopper ash by TSA

Activation			Sample designation for different molarity of the NaOH						
T (h)	Step	T_c (h)	M = 0.5	1.0	1.5	2.0	2.5	3.0	3.5
12	PT	12	0.5-PT-12	1.0-PT-12	1.5-PT-12	2.0-PT-12	2.5-PT-12	3.0-PT-12	3.5-PT-12
	R1	24	0.5-R1-12	1.0-R1-12	1.5-R1-12	2.0-R1-12	2.5-R1-12	3.0-R1-12	3.5-R1-12
	R2	36	0.5-R2-12	1.0-R2-12	1.5-R2-12	2.0-R2-12	2.5-R2-12	3.0-R2-12	3.5-R2-12
24	PT	24	0.5-PT-24	1.0-PT-24	1.5-PT-24	2.0-PT-24	2.5-PT-24	3.0-PT-24	3.5-PT-24
	R1	48	0.5-R1-24	1.0-R1-24	1.5-R1-24	2.0-R1-24	2.5-R1-24	3.0-R1-24	3.5-R1-24
	R2	72	0.5-R2-24	1.0-R2-24	1.5-R2-24	2.0-R2-24	2.5-R2-24	3.0-R2-24	3.5-R2-24
36	PT	36	0.5-PT-36	1.0-PT-36	1.5-PT-36	2.0-PT-36	2.5-PT-36	3.0-PT-36	3.5-PT-36
	R1	72	0.5-R1-36	1.0-R1-36	1.5-R1-36	2.0-R1-36	2.5-R1-36	3.0-R1-36	3.5-R1-36
	R2	108	0.5-R2-36	1.0-R2-36	1.5-R2-36	2.0-R2-36	2.5-R2-36	3.0-R2-36	3.5-R2-36
48	PT	48	0.5-PT-48	1.0-PT-48	1.5-PT-48	2.0-PT-48	2.5-PT-48	3.0-PT-48	3.5-PT-48
	R1	96	0.5-R1-48	1.0-R1-48	1.5-R1-48	2.0-R1-48	2.5-R1-48	3.0-R1-48	3.5-R1-48
	R2	144	0.5-R2-48	1.0-R2-48	1.5-R2-48	2.0-R2-48	2.5-R2-48	3.0-R2-48	3.5-R2-48

Table 5.2 Sample designation for conventionally treated ash

M	*T* (h)							
	12	24	36	48	12	24	36	48
0.5	ALA1	ALA2	ALA3	ALA4	AHA1	AHA2	AHA3	AHA4
1.0	ALA5	ALA6	ALA7	ALA8	AHA5	AHA6	AHA7	AHA8
2.0	ALA9	ALA10	ALA11	ALA12	AHA9	AHA10	AHA11	AHA12
3.5	ALA13	ALA14	ALA15	ALA16	AHA13	AHA14	AHA15	AHA16

5.2.2 Three Step Activation by Fusion Technique

Researchers have confirmed that fusion of the fly ash results in loss of its volatile matter and unburnt carbon, which can be quantified in terms of net loss in weight of the fused ash [18]. Furthermore, fusion of a mixture of fly ash and NaOH has been widely explored [19–21], by resorting to its continuous heating for longer time at high temperature. Accordingly, they have established two main critical attributes as: (i) optimum temperature (=500 °C) and (ii) fusion period (=2 h). In addition, researchers have highlighted problems of inappropriate contact between the NaOH and fly ash in dry phase, hence incomplete interaction between them. In such a situation, they have demonstrated that zeolitization (i.e., nucleation and growth of zeolite Na-X)of fused fly ash (i.e., enriched with sodium alumino silicate) can be initiated by its hydrothermal curing. In fact, hydrothermal curing produces substantial quantity of alkaline liquid by-product, which, in general, is disposed of in the environment, resulting in its degradation. To overcome this, a suitably modified fusion process without any hydrothermal curing, which facilitates sufficient contact between the ash and NaOH, appears to be a panacea. In this context, usage of a small quantity of water (e.g., 5 ml) has been preferred to dissolve the NaOH pellets, before mixing it with the fly ash. Thus by inclusion of such a simple modification in the conventional fusion process, the fly ash particle can be coated with a thin layer of NaOH, before start of the actual fusion process. On the other hand, the fusion results in creation of a thin zeolitic (i.e., silicates) layer around the residual fly ash particles. Consequently, like Step-1 of the TSA by hydrothermal technique (as discussed above), further conversion of the fly ash to silicate like compounds, gets restricted. As such, complete fusion of the fly ash with NaOH (referred as three step activation, TSA, by fusion technique) has to beensured; though detaching the silicate layer from the inactivated portion of the fly ash particle is a challenging task [11]. A typical flow chart of the entire process of the TSA by fusion has been presented in Fig. 5.3.

It can be noticed that a manual grinding of a fusion product from Step-1 (i.e., primary fusion, designated by F1) precedes Step-2 (i.e., secondary fusion, F2) and Step-3 (i.e., tertiary fusion, F3). In order to study the major characteristics (viz., weight loss, specific gravity, chemical composition, cation exchange capacity, mineralogy, morphology, infrared absorption and variation of Si and Al atoms in the structure of the FAZ) of the products of the fusions, F1 to F3, a comparative

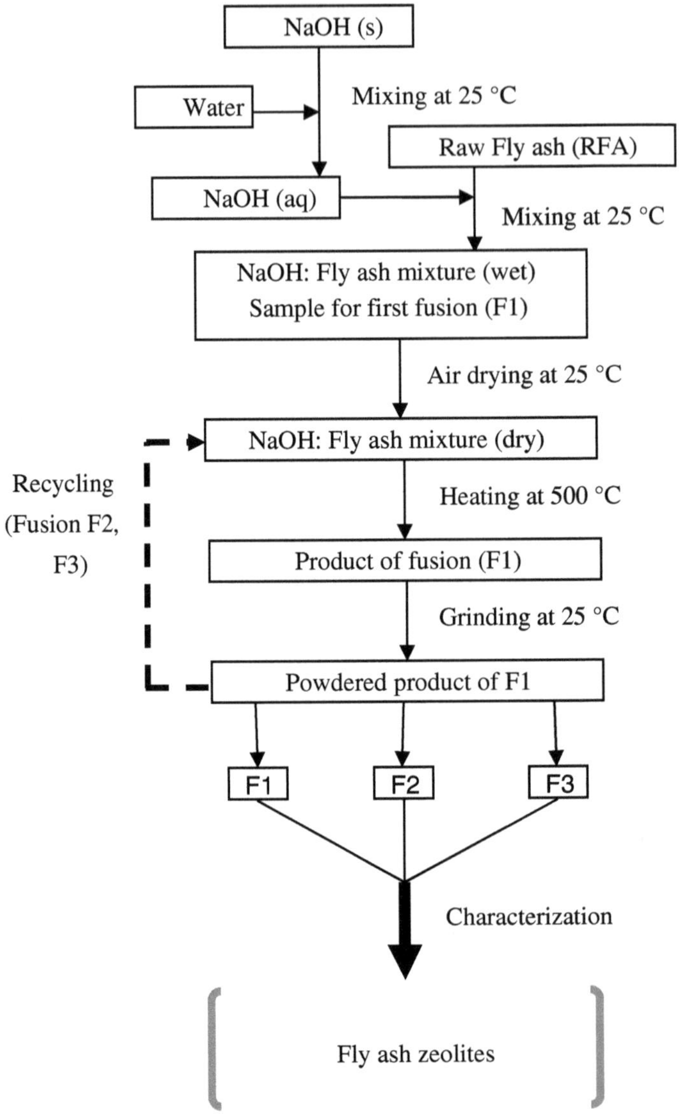

Fig. 5.3 A typical process flowchart for three-step activation of the fly ash by fusion

Table 5.3 Sample designation for the products from three-step fusion

Temp (°C)	Time (h)	Step	T_c (h)	Sample designation for different NaOH/Fly ash ratios						
				0.2	0.4	0.6	0.8	1.0	1.2	1.4
500	2	F1	2	0.2-F1	0.4-F1	0.6-F1	0.8-F1	1.0-F1	1.2-F1	1.4-F1
500	2	F2	4	0.2-F2	0.4-F2	0.6-F2	0.8-F2	1.0-F2	1.2-F2	1.4-F2
500	2	F3	6	0.2-F3	0.4-F3	0.6-F3	0.8-F3	1.0-F3	1.2-F3	1.4-F3

investigation of the test data is essential. In this context, the fusion products have been assigned designations as listed in Table 5.3.

5.3 Characterization of Products

Based on the variation in the reaction conditions, the quantification of Si and Al concentrations in the supernatant of each step of the TSA by hydrothermal techniques has been given significance. Also, the final product obtained from each of the experiments, may contain varying proportions of fly ash zeolites. As such, it is of utmost significance to completely characterize the final product for various parameters [19, 22–25].

5.3.1 *Characterization of the Supernatant*

A representative sample of the supernatant (ST) was obtained after each step of the TSA by hydrothermal technique and centrifugation (at $\approx$2000 rpm). The sample was filtered by using Whatman filter paper No. 42 and diluted sufficiently, in the subsequent step, for analysis [14].

5.3.1.1 Determination of *pH* and Electrical Conductivity (*EC*)

The *pH* and *EC* of the sample were measured by using a water quality analyzer (Model PE 136, Elico Ltd., India). The results of the *pH* and *EC* tests are presented in Tables 5.4, 5.5, 5.6 and 5.7, respectively [14].

5.3.1.2 Determination of Si and Al Contents

A clean and filtered sample of the supernatant, ST, was used for determination of concentration of soluble Si and Al, present in the fly ash.

Table 5.4 *pH* of the supernatant obtained from TSA of the fly ash for 12 and 24 h

Sample	*pH*				Sample	*pH*			
	Trial			Avg.		Trial			Avg.
	1	2	3			1	2	3	
0.5-PT-12	10	10.4	10.5	10.3	0.5-PT-24	12	11.9	11.8	11.9
1.0-PT-12	11.9	12.5	12.5	12.3	1.0-PT-24	12.2	12.4	12.3	12.3
1.5-PT-12	12.3	11.8	11.9	12.0	1.5-PT-24	12.4	12.5	12.3	12.4
2.0-PT-12	12.5	12.8	12.5	12.6	2.0-PT-24	12.4	12.6	12.5	12.5
2.5-PT-12	12.5	12.8	12.8	12.7	2.5-PT-24	12.7	12.6	12.5	12.6
3.0-PT-12	12.3	12.6	12.6	12.5	3.0-PT-24	12.6	12.6	12.6	12.6
3.5-PT-12	12.5	12.8	12.5	12.6	3.5-PT-24	12.6	12.8	12.7	12.7
0.5-R1-12	11.6	11.5	11.1	11.4	0.5-R1-24	10.8	10.7	10.9	10.8
1.0-R1-12	12.3	12.6	12.6	12.5	1.0-R1-24	11.8	11.9	12	11.9
1.5-R1-12	12	12.5	12.4	12.3	1.5-R1-24	12.3	12.5	12.4	12.4
2.0-R1-12	12.3	12.6	12.6	12.5	2.0-R1-24	12.6	12.5	12.7	12.6
2.5-R1-12	12.6	12.7	12.5	12.6	2.5-R1-24	12.8	12.8	12.8	12.8
3.0-R1-12	12.7	12.5	12.6	12.6	3.0-R1-24	12.9	12.8	13	12.9
3.5-R1-12	12.7	13	13	12.9	3.5-R1-24	12.9	13.1	13	13
0.5-R2-12	11.1	10.9	11	11.0	0.5-R2-24	10	9.9	9.8	9.9
1.0-R2-12	12.6	12.4	12.5	12.5	1.0-R2-24	11.3	11.5	11.4	11.4
1.5-R2-12	12.3	12.5	12.4	12.4	1.5-R2-24	12.4	12.5	12	12.3
2.0-R2-12	12.6	12.8	12.7	12.7	2.0-R2-24	12.4	12.3	12.2	12.3
2.5-R2-12	12.8	12.6	12.7	12.7	2.5-R2-24	12.5	12.4	12.6	12.5
3.0-R2-12	13	12.8	12.6	12.8	3.0-R2-24	12.7	12.7	12.7	12.7
3.5-R2-12	12.9	12.8	13	12.9	3.5-R2-24	13	12.9	13.1	13

For this purpose, the inductively coupled plasma-atomic emission spectrometry (ICP-AES) set up (Model 8440, Plasma lab, GBC Scientific Equipment Pvt. Ltd., Australia) was employed. The results of the analyses are presented in Tables 5.8, 5.9, 5.10 and 5.11, respectively.

Based on reanalysis of ICP-AES data, percentage variation of these elements has been quantified by assigning a parameter, α, defined as per the relationship presented in Eq. 5.1.

$$\alpha = (C \cdot V/p_1 \cdot \mathrm{m}) \cdot 10^{-4} \quad (5.1)$$

where, C is the concentration of either Si^{4+} or Al^{3+} (in ppm) in the supernatant, V is the volume (in ml) of the ST, used for the activation, p_1 is the percentage of Si and Al in oxide form, present in the residue, before treatment, respectively, and m is the mass (in g) of the residue.

Table 5.5 *pH* of the supernatant obtained from TSA of the fly ash for 36 and 48 h

Sample	*pH*				Sample	*pH*			
	Trial			Avg.		Trial			Avg.
	1	2	3			1	2	3	
0.5-PT-36	12.8	12.7	12.6	12.7	0.5-PT-48	13.41	13.42	13.43	13.42
1.0-PT-36	13.1	13.2	13.3	13.2	1.0-PT-48	13.7	13.71	13.72	13.71
1.5-PT-36	13.45	13.47	13.46	13.46	1.5-PT-48	13.88	13.9	13.83	13.87
2.0-PT-36	13.5	13.5	13.5	13.5	2.0-PT-48	13.34	13.34	13.34	13.34
2.5-PT-36	13.65	13.7	13.63	13.66	2.5-PT-48	13.52	13.51	13.53	13.52
3.0-PT-36	13.9	13.87	13.87	13.88	3.0-PT-48	13.63	13.63	13.63	13.63
3.5-PT-36	13.94	13.95	13.9	13.93	3.5-PT-48	13.74	13.8	13.68	13.74
0.5-R1-36	11.59	11.61	11.6	11.6	0.5-R1-48	12.74	12.75	12.73	12.74
1.0-R1-36	12.74	12.73	12.78	12.75	1.0-R1-48	13.43	13.44	13.42	13.43
1.5-R1-36	13.22	13.23	13.21	13.22	1.5-R1-48	13.69	13.7	13.68	13.69
2.0-R1-36	13.36	13.38	13.37	13.37	2.0-R1-48	13.25	13.3	13.23	13.26
2.5-R1-36	12.6	12.7	12.8	12.7	2.5-R1-48	13.62	13.61	13.63	13.62
3.0-R1-36	13.84	13.86	13.82	13.84	3.0-R1-48	13.53	13.54	13.52	13.53
3.5-R1-36	14.32	14.3	14.34	14.32	3.5-R1-48	13.63	13.64	13.62	13.63
0.5-R2-36	10.96	10.9	11.02	10.96	0.5-R2-48	11.32	11.3	11.34	11.32
1.0-R2-36	11.4	11.3	11.5	11.4	1.0-R2-48	11.95	12	12.02	11.99
1.5-R2-36	12	12.1	11.9	12	1.5-R2-48	12.6	12.5	12.82	12.64
2.0-R2-36	12.72	12.73	12.74	12.73	2.0-R2-48	12.5	12.49	12.51	12.5
2.5-R2-36	11.89	11.9	11.88	11.89	2.5-R2-48	12.48	12.49	12.44	12.47
3.0-R2-36	14.1	14	13.9	14	3.0-R2-48	13.28	13.3	13.26	13.28
3.5-R2-36	13.97	13.99	13.98	13.98	3.5-R2-48	13.45	13.5	13.49	13.48

5.3.2 *Characterization of the Alkali Activated Fly Ash (AAF)*

It is of utmost significance to completely characterize the product obtained after each step of the TSA for various parameters viz., specific gravity, specific surface area, chemical composition, cation exchange capacity, mineralogical composition, crystallinity, morphology, infrared transmittance and acid soluble content [19, 22–25].

5.3.2.1 Determination of Specific Gravity

The specific gravity, *G*, of the sample was determined by employing Ultra Pycnometer (Quanta chrome, USA), which works on the principle of adsorption of Helium gas (as per, ASTM D 5550-00) [26]. For the sake of accuracy the average of the three values of *G* has been presented in Tables 5.12, 5.13 and 5.14 for TSA by hydrothermal and fusion products, respectively.

Table 5.6 *EC* of the supernatant obtained from TSA of the fly ash for 12 and 24 h

Sample	*EC* (μS/cm)				Sample	*EC* (μSμ/cm)			
	Trial			Avg.		Trial			Avg.
	1	2	3			1	2	3	
0.5-PT-12	28	27.6	26.9	27.5	0.5-PT-24	26	25.5	25.99	25.83
1.0-PT-12	132.3	132.2	132.4	132.3	1.0-PT-24	76	77	76.26	76.42
1.5-PT-12	75.4	75.3	75.2	75.3	1.5-PT-24	120	119.5	117.5	119
2.0-PT-12	124.5	124.6	124.7	124.6	2.0-PT-24	168	167	168.4	167.8
2.5-PT-12	156.4	156.5	156.3	156.4	2.5-PT-24	184.5	184.5	184.8	184.6
3.0-PT-12	263.5	264	263	263.5	3.0-PT-24	217.5	217.4	217.3	217.4
3.5-PT-12	288	287.5	287	287.5	3.5-PT-24	264	265.9	260.9	263.6
0.5-R1-12	23.6	24	23.2	23.6	0.5-R1-24	21.1	21.21	21.2	21.17
1.0-R1-12	109.5	110	109.3	109.6	1.0-R1-24	54.6	54.62	54.64	54.62
1.5-R1-12	125	126	125.8	125.6	1.5-R1-24	91.65	91.7	91.6	91.65
2.0-R1-12	169	168	169.4	168.8	2.0-R1-24	125.5	125	125.7	125.4
2.5-R1-12	195	194	194.5	194.5	2.5-R1-24	148	146	148.5	147.5
3.0-R1-12	226.5	227	226	226.5	3.0-R1-24	170	169	169.8	169.6
3.5-R1-12	245	247	246.9	246.3	3.5-R1-24	216	215	214	215
0.5-R2-12	23	22	22.2	22.4	0.5-R2-24	20	21	20.2	20.4
1.0-R2-12	75.3	75.5	75.4	75.4	1.0-R2-24	33.69	33.7	33.56	33.65
1.5-R2-12	90.5	91	89.4	90.3	1.5-R2-24	67	66	66.26	66.42
2.0-R2-12	126.9	126.8	126.7	126.8	2.0-R2-24	91.64	91.7	91.58	91.64
2.5-R2-12	146.1	147	145.5	146.2	2.5-R2-24	91.35	92	90.7	91.35
3.0-R2-12	172.8	175	170.6	172.8	3.0-R2-24	99	100	95.39	98.13
3.5-R2-12	205	206	205.8	205.6	3.5-R2-24	129.4	129.4	129.4	129.4

5.3.2.2 Determination of Specific Surface Area

Specific surface area of the RFA and the AAF was determined by following two methods: 1. ethylene glycol monoethyl ether (EGME) method and 2. Brunauer-Emmett-Teller (BET) method. A detail description of these methods is presented in the following sections.

EGME Method

The specific surface area (*SSA*) of the sample has been determined by EGME method, which is based on the principle of adsorption of the EGME by the sample.

The adsorption process has been started by saturating the test sample with P_2O_5, equilibrating it in vacuum over $CaCl_2$-EGME solvate and weighing to find the equilibrium state. The *SSA* can then be determined from the mass of the EGME retained by the sample in comparison to the amount retained by pure

Table 5.7 *EC* of the supernatant obtained from TSA of the fly ash for 36 and 48 h

Sample	*EC* (μS/cm)				Sample	*EC* (μS/cm)			
	Trial			Avg.		Trial			Avg.
	1	2	3			1	2	3	
0.5-PT-36	29.35	29.5	29.11	29.32	0.5-PT-48	46.5	46.6	46.07	46.39
1.0-PT-36	86.7	87	86.46	86.72	1.0-PT-48	94	94.2	93.56	93.92
1.5-PT-36	123	124	123.2	123.4	1.5-PT-48	118	117.9	118.1	118
2.0-PT-36	148	149.5	148	148.5	2.0-PT-48	154.5	154	154.1	154.2
2.5-PT-36	212.5	211.9	212.2	212.2	2.5-PT-48	199.8	199	201.2	200
3.0-PT-36	214	213	213.5	213.5	3.0-PT-48	227.3	226.9	226.8	227
3.5-PT-36	234	234.5	233.5	234	3.5-PT-48	232.5	232.2	232.5	232.4
0.5-R1-36	20.08	20	20.16	20.08	0.5-R1-48	20.9	20.6	20.93	20.81
1.0-R1-36	53	52	52.11	52.37	1.0-R1-48	61.05	61	61.13	61.06
1.5-R1-36	89.38	90	89	89.46	1.5-R1-48	88.4	8	168.2	88.2
2.0-R1-36	109	108	109.4	108.8	2.0-R1-48	108.5	108	108.4	108.3
2.5-R1-36	148.5	148	150.2	148.9	2.5-R1-48	111.1	111.4	111.4	111.3
3.0-R1-36	152.6	152	152.6	152.4	3.0-R1-48	156	155.95	156.95	156.3
3.5-R1-36	177.9	177	177.6	177.5	3.5-R1-48	170.98	171.3	171.92	171.4
0.5-R2-36	21	20.8	20.39	20.73	0.5-R2-48	19.3	19.5	18.98	19.26
1.0-R2-36	32	33	32.68	32.56	1.0-R2-48	33.2	33	33.1	33.1
1.5-R2-36	54	53.5	53.36	53.62	1.5-R2-48	60.6	60.4	60.74	60.58
2.0-R2-36	67.7	68	67.4	67.7	2.0-R2-48	64	63.6	63.38	63.66
2.5-R2-36	19.4	19.6	19.14	19.38	2.5-R2-48	60.48	60.45	60.27	60.4
3.0-R2-36	101.9	102	100.9	101.6	3.0-R2-48	89	89.6	90.56	89.72
3.5-R2-36	89	90	87.91	88.97	3.5-R2-48	88.53	88.48	89.03	88.68

Montmorillonite clay, which has been assumed to have a SSA_{EGME} of 810 m^2/g [27, 28], by employing Eq. 5.2

$$SSA_{EGME} = W_a/(0.000286 \times W_{AFA}) \tag{5.2}$$

where, W_a is the weight of the EGME retained by the sample (in g), which is equal to the final slurry weight minus the weight of the sample, W_{AFA}. 0.000286 corresponds to the weight of the EGME required to form a mono-molecular layer on a square meter of surface of the sample. The value of the SSA_{EGME} for different samples is presented in Tables 5.15, 5.16, 5.17 and 5.18.

BET Method

A Brunauer-Emmett-Teller (BET) set up (ASAP 2020 system V 3.01 H fitted with Micromeritics patented isothermal jackets, Norcross, GA 30093-1877) [29] was used and a set of standard procedures were followed for this purpose [30]. The

Table 5.8 Si content of the supernatant from TSA of the fly ash for 12 and 24 h

Sample	Si (ppm)				Sample	Si (ppm)			
	Trial			Avg.		Trial			Avg.
	1	2	3			1	2	3	
0.5-PT-12	3200	3256	3006	3154	0.5-PT-24	6050	6900	4348	5766
1.0-PT-12	4948	5020	4408	4792	1.0-PT-24	5000	5500	5175	5225
1.5-PT-12	6000	5920	5408	5776	1.5-PT-24	4245	4100	4090	4145
2.0-PT-12	4800	4670	4138	4536	2.0-PT-24	5200	5000	5388	5196
2.5-PT-12	7000	7560	7208	7256	2.5-PT-24	11,100	10,800	10,947	10,949
3.0-PT-12	6960	7000	6698	6886	3.0-PT-24	12,000	11,500	11,996	11,832
3.5-PT-12	3535	4000	4426	3987	3.5-PT-24	23,600	23,500	23,160	23,420
0.5-R1-12	7600	6800	7557	7319	0.5-R1-24	11,600	10,960	11,088	11,216
1.0-R1-12	105	98	102.46	101.82	1.0-R1-24	9590	10,000	9165	9585
1.5-R1-12	6900	6700	6905	6835	1.5-R1-24	11,350	11,600	10,800	11,250
2.0-R1-12	10,900	11,500	11,194	11,198	2.0-R1-24	12,000	11,500	12,050	11,850
2.5-R1-12	10,200	10,800	11,709	10,903	2.5-R1-24	12,500	10,900	12,000	11,800
3.0-R1-12	8900	9000	8893	8931	3.0-R1-24	11,000	11,500	10,635	11,045
3.5-R1-12	12,000	15,900	13,833	13,911	3.5-R1-24	10,350	9900	10,380	10,210
0.5-R2-12	21,500	20,500	21,270	21,090	0.5-R2-24	26,900	25,800	24,715	25,805
1.0-R2-12	9800	9950	10,082	9944	1.0-R2-24	7055	8000	7670	7575
1.5-R2-12	16,800	15,050	13,732	15,194	1.5-R2-24	9785	10,000	9300	9695
2.0-R2-12	200	250	164.1	204.7	2.0-R2-24	15,900	16,050	15,720	15,890
2.5-R2-12	110	99	116.26	108.42	2.5-R2-24	30,500	29,500	32,100	30,700
3.0-R2-12	13,050	10,900	14,714	12,888	3.0-R2-24	8950	9300	8825	9025
3.5-R2-12	20,000	21,090	19,783	20,291	3.5-R2-24	15,950	18,500	17,825	17,425

analysis temperature was maintained as that of liquid N_2 (i.e. −195.489 °C) and sample (maximum 1 g) was degassed automatically under vacuum (5 × 10^{-3} mmHg). The results such as *SSA*, pore volume and types of pore sizes (micro-pores below 2 nm, meso-pores between 2 and 50 nm and macro-pores above 50 nm) are presented in following sections of this chapter.

5.3.2.3 Calculation for Weight Loss in Fused Residues

The weight loss of the products of the TSA by fusion technique was calculated by following relationship as described in Eq. 5.3.

$$\text{Weight loss (\%)} = \frac{W_i - W_f}{W_i} \times 100 \tag{5.3}$$

where, W_i and W_f are the initial and final weights of the sample (in g) and presented in Table 5.19.

Table 5.9 Si content of the supernatant from TSA of the fly ash for 36 and 48 h

Sample	Si (ppm)				Sample	Si (ppm)			
	Trial			Avg.		Trial			Avg.
	1	2	3			1	2	3	
0.5-PT-36	5500	5000	3945	4815	0.5-PT-48	29,050	28,500	27,752	28,434
1.0-PT-36	6900	6850	6872	6874	1.0-PT-48	14,950	15,300	15,551	15,267
1.5-PT-36	7010	6950	6953	6971	1.5-PT-48	42,500	41,950	41,613	42,021
2.0-PT-36	10,600	11,000	10,065	10,555	2.0-PT-48	11,300	12,100	10,650	11,350
2.5-PT-36	55,050	55,500	52,014	54,188	2.5-PT-48	4895	4900	4950	4915
3.0-PT-36	5450	5500	5154	5368	3.0-PT-48	8050	7580	7623	7751
3.5-PT-36	35,000	33,000	33,592	33,864	3.5-PT-48	7840	7900	7792	7844
0.5-R1-36	6250	7050	5015	6105	0.5-R1-48	2050	2200	2290	2180
1.0-R1-36	10,000	9900	9875	9925	1.0-R1-48	4400	4555	4065	4340
1.5-R1-36	13,500	14,000	12,454	13,318	1.5-R1-48	6540	6600	6180	6440
2.0-R1-36	16,082	16,525	15,939	16,182	2.0-R1-48	26,540	25,500	25,261	25,767
2.5-R1-36	11,900	11,500	12,261	11,887	2.5-R1-48	5714	5700	5728	5714
3.0-R1-36	38,900	36,900	38,605	38,135	3.0-R1-48	16,300	16,800	16,040	16,380
3.5-R1-36	55,950	54,850	57,365	56,055	3.5-R1-48	12,800	11,900	11,990	12,230
0.5-R2-36	6390	5500	6926	6272	0.5-R2-48	3100	2890	2740	2910
1.0-R2-36	16,300	16,100	15,807	16,069	1.0-R2-48	6900	6800	6820	6840
1.5-R2-36	60,200	59,900	63,818	61,306	1.5-R2-48	10,500	10,500	9090	10,030
2.0-R2-36	24,900	23,050	23,690	23,880	2.0-R2-48	13,600	12,400	12,907	12,969
2.5-R2-36	8950	8500	9105	8852	2.5-R2-48	21,450	20,500	20,957	20,969
3.0-R2-36	29,060	30,000	28,127	29,062	3.0-R2-48	15,000	14,200	15,092	14,764
3.5-R2-36	39,500	39,200	38,915	39,500	3.5-R2-48	17,050	16,500	15,473	16,341

5.3.2.4 Determination of Chemical Composition

The chemical composition of the sample in the form of major oxides was determined using an X-ray fluorescence (XRF) set up (Philips 1410, Holland). The details of sample preparation and calibration of instruments are presented in the following.

A uniform mixture of 4 g of the sample and 1 g of micro-crystalline cellulose was prepared and thoroughly mixed with few drops of isopropyl alcohol. The wet mixture was exposed to infrared lamp for slow drying. The dried mixture was again uniformly mixed. An aluminium sample holder (with inner diameter 33 mm and height 12 mm) was filled up to two third with same cellulose powder at bottom followed by the dried sample with uniform spread and leveled surface. The filled up sample holder is thereafter compressed in a hydraulic jack up to 15 ton pressure. The XRF set up was calibrated for physical and chemical calibration to eliminate the noise region pulses. The angle calibration was performed to configure the set up for measuring the peak height and background points on either side of the peak at specified angle. This calibration was done to eliminate the error involved due to

Table 5.10 Al content of the supernatant from TSA of the fly ash for 12 and 24 h

Sample	Al (ppm)				Sample	Al (ppm)			
	Trial			Avg.		Trial			Avg.
	1	2	3			1	2	3	
0.5-PT-12	23.5	24.2	21	22.9	0.5-PT-24	2.9	3.4	3.45	3.25
1.0-PT-12	2	1.5	2.2	1.9	1.0-PT-24	0.5	0.48	0.37	0.45
1.5-PT-12	0.95	1	1.05	1	1.5-PT-24	0.75	0.85	0.8	0.8
2.0-PT-12	100.5	101.6	103.6	101.9	2.0-PT-24	10.5	9	8.1	9.2
2.5-PT-12	16.5	17	18.4	17.3	2.5-PT-24	7	6.5	6.6	6.7
3.0-PT-12	15	14.2	13.7	14.3	3.0-PT-24	11.8	10.6	12.1	11.5
3.5-PT-12	22.6	21.8	24.3	22.9	3.5-PT-24	58	56.8	56.65	57.15
0.5-R1-12	2.5	3	3.5	3	0.5-R1-24	24.9	25	26.9	25.6
1.0-R1-12	81	74	67	74	1.0-R1-24	16.5	15.95	17.2	16.55
1.5-R1-12	101.5	100.8	103.1	101.8	1.5-R1-24	47.5	46.9	47.65	47.35
2.0-R1-12	79.9	77.5	82.3	79.9	2.0-R1-24	45.1	39.5	37.05	40.55
2.5-R1-12	72.5	72.4	71.1	72	2.5-R1-24	40.5	39.9	40.2	40.2
3.0-R1-12	33	35	34	34	3.0-R1-24	80	78	78.7	78.9
3.5-R1-12	35	37	36	36	3.5-R1-24	59	57.5	58.4	58.3
0.5-R2-12	73	75	74	74	0.5-R2-24	66.3	64.9	66.05	65.75
1.0-R2-12	62	63	64	63	1.0-R2-24	82	91	82.15	85.05
1.5-R2-12	56	57	58	57	1.5-R2-24	87	86.7	87.45	87.05
2.0-R2-12	86	84	85	85	2.0-R2-24	120	119	122.2	120.4
2.5-R2-12	72	74	76	74	2.5-R2-24	120.5	119	118.25	119.25
3.0-R2-12	85	86	87	86	3.0-R2-24	105	104	108.4	105.8
3.5-R2-12	20	22	21	21	3.5-R2-24	185	186.6	182.95	184.85

uneven and slanting base line. Standard reference material (SRM) supplied by UGCS, USA was used for the calibration of the instruments. Later, the chemical calibration of the instrument was performed with the help of matrix matched with international standard reference material [Fly ash (2689, 2690, 2691), cements (354, 372, 372/1) and soils (SO-2, SO-3, SO-4)], procured from the NIST, USA. The chemical composition of the sample was determined by mounting the compressed dish (pellet) in the sample holder of the XRF set up. The obtained results are presented in Tables 5.20, 5.21, 5.22, 5.23 and 5.24 for the hydrothermal and fusion products, respectively.

Based on reanalysis of data from these tables, percentage variation (by mass) of major oxides (SiO_2 and Al_2O_3), due to each step of activation, was calculated by defining a relationship, as described in Eq. 5.4.

$$\gamma = 100 \cdot (q_2 - q_1)/q_1 \tag{5.4}$$

Table 5.11 Al content of the supernatant from TSA of the fly ash for 36 and 48 h

Sample	Al (ppm)				Sample	Al (ppm)			
	Trial			Avg.		Trial			Avg.
	1	2	3			1	2	3	
0.5-PT-36	6.3	6	5.7	6	0.5-PT-48	570	550	560	560
1.0-PT-36	38	37.5	35.5	37	1.0-PT-48	225	228	219	224
1.5-PT-36	27	28	29	28	1.5-PT-48	1009	1029	1112	1050
2.0-PT-36	46	46.5	45.5	46	2.0-PT-48	68	60	64	64
2.5-PT-36	304	300	299	301	2.5-PT-48	100	105	95	100
3.0-PT-36	179	178.8	179.2	179	3.0-PT-48	190	193	193	192
3.5-PT-36	274	275	276	275	3.5-PT-48	194	196	195	195
0.5-R1-36	5.5	4.5	5	5	0.5-R1-48	11	10.5	8.5	10
1.0-R1-36	10	10	10	10	1.0-R1-48	40.5	39.8	39.7	40
1.5-R1-36	19.5	21	19.5	20	1.5-R1-48	69.5	70.6	69.9	70
2.0-R1-36	46	45	44	45	2.0-R1-48	229	235	223	229
2.5-R1-36	92	92	92	92	2.5-R1-48	31	30.9	34.1	32
3.0-R1-36	117	116.5	117.5	117	3.0-R1-48	188	189	187	188
3.5-R1-36	209	209.8	211.2	210	3.5-R1-48	27	26	28	27
0.5-R2-36	12.2	13.5	10.3	12	0.5-R2-48	20	21	19	20
1.0-R2-36	43	45	41	43	1.0-R2-48	10	10	10	10
1.5-R2-36	45	43.8	43.2	44	1.5-R2-48	20	19.8	20.2	20
2.0-R2-36	35	36	37	36	2.0-R2-48	115	114	113	114
2.5-R2-36	187	186	185	186	2.5-R2-48	69.6	70	67.4	69
3.0-R2-36	69	68	70	69	3.0-R2-48	38	39	40	39
3.5-R2-36	250	249	248	249	3.5-R2-48	77.6	78	78.4	78

where, $q_1 = p_1 \cdot m$ and $q_2 = p_2 \cdot m$, q_1 and q_2 are quantities (in g) of either SiO_2 or Al_2O_3, corresponding to p_1 and p_2, the percentages of these oxides in the residue, before and after a treatment, respectively, and *m* is the mass (in g) of the residue.

The relationship between γ and cumulative activation duration, T_c, has been depicted in following section.

5.3.2.5 Determination of Cation Exchange Capacity (*CEC*)

The *CEC* of a material can be determined by using the most common method [31–34] as described below.

The cation exchange capacity of a zeolitic product can be determined by mixing 1.066 g of powder sample with 9 ml of CH_3COONa solution in a mechanical shaker for 5 min. The mixture was centrifuged until a clear supernatant is obtained and later it is decanted. The similar process can be repeated for another two times to ensure that all cations in the zeolite have been replaced with Na^+. The Na^+ laden zeolites were mixed with 9 ml of 99 % isopropyl alcohol and shaken in a

Table 5.12 Specific gravity (*G*) of the AAF from TSA for 12 and 24 h

Sample	*G*				Sample	*G*			
	Trial			Avg.		Trial			Avg.
	1	2	3			1	2	3	
0.5-PT-12	2.6	2.58	2.65	2.61	0.5-PT-24	2.52	2.56	2.57	2.55
1.0-PT-12	2.39	2.41	2.4	2.40	1.0-PT-24	2.55	2.57	2.59	2.57
1.5-PT-12	2.59	2.67	2.72	2.66	1.5-PT-24	2.95	2.96	2.91	2.94
2.0-PT-12	2.6	2.59	2.55	2.58	2.0-PT-24	2.66	2.58	2.71	2.65
2.5-PT-12	2.5	2.52	2.51	2.51	2.5-PT-24	2.54	2.57	2.57	2.56
3.0-PT-12	2.59	2.58	2.63	2.60	3.0-PT-24	2.6	2.58	2.65	2.61
3.5-PT-12	2.39	2.4	2.41	2.40	3.5-PT-24	2.56	2.55	2.57	2.56
0.5-R1-12	2.62	2.63	2.64	2.63	0.5-R1-24	2.55	2.56	2.51	2.54
1.0-R1-12	2.54	2.55	2.5	2.53	1.0-R1-24	2.52	2.52	2.52	2.52
1.5-R1-12	2.65	2.59	2.56	2.60	1.5-R1-24	2.6	2.64	2.65	2.63
2.0-R1-12	2.66	2.56	2.73	2.65	2.0-R1-24	2.63	2.59	2.61	2.61
2.5-R1-12	2.52	2.5	2.51	2.51	2.5-R1-24	2.5	2.48	2.46	2.48
3.0-R1-12	2.48	2.48	2.51	2.49	3.0-R1-24	2.61	2.58	2.61	2.60
3.5-R1-12	2.58	2.57	2.56	2.57	3.5-R1-24	2.55	2.53	2.6	2.56
0.5-R2-12	2.43	2.45	2.44	2.44	0.5-R2-24	2.55	2.57	2.53	2.55
1.0-R2-12	2.48	2.45	2.48	2.47	1.0-R2-24	2.48	2.51	2.54	2.51
1.5-R2-12	2.58	2.57	2.56	2.57	1.5-R2-24	2.54	2.55	2.56	2.55
2.0-R2-12	2.58	2.57	2.56	2.57	2.0-R2-24	2.52	2.54	2.5	2.52
2.5-R2-12	2.5	2.51	2.37	2.46	2.5-R2-24	2.29	2.39	2.34	2.34
3.0-R2-12	2.53	2.57	2.67	2.59	3.0-R2-24	2.55	2.5	2.45	2.50
3.5-R2-12	2.54	2.56	2.49	2.53	3.5-R2-24	2.6	2.62	2.61	2.61
SZP	2.13	2.15	2.14	2.14	RFA	2.18	2.17	2.19	2.18

mechanical shaker for 5 min to insure removal of all organic matters. The resulting supernatant after centrifugation can be decanted and the same process repeated two times to ensure that the zeolite becomes clean and laden with Na^+ only. Later, the residue was mixed with 9 ml of CH_3COONH_4 solution and the similar process was repeated two times to ensure that all the Na^+ have been replaced by the NH_4^+. The supernatants of all the three last centrifugations were collected in a volumetric flask of certain capacity, *V*, and diluted to 100 ml by adding ammonium acetate solution. The concentration of Na^+ in the supernatant was determined with the help of ICP-AES. The sodium binding capacity of such samples was computed by using the Eq. 5.5.

$$CEC = \frac{C \times V \times 100}{1000 \times m} \tag{5.5}$$

where, *C* is the concentration of Na^+ (meq./l) in the solution obtained after last three washings, denoted as epm (i.e. equivalent parts per million i.e. parts per million,

Table 5.13 Specific gravity (G) of the AAF from TSA for 36 and 48 h

Sample	G				Sample	G			
	Trial			Avg.		Trial			Avg.
	1	2	3			1	2	3	
0.5-PT-36	2.65	2.65	2.66	2.65	0.5-PT-48	2.46	2.48	2.48	2.48
1.0-PT-36	2.72	2.73	2.75	2.73	1.0-PT-48	2.51	2.53	2.53	2.52
1.5-PT-36	2.67	2.67	2.67	2.67	1.5-PT-48	2.58	2.58	2.58	2.58
2.0-PT-36	2.70	2.72	2.72	2.71	2.0-PT-48	2.55	2.57	2.58	2.57
2.5-PT-36	2.59	2.61	2.6	2.6	2.5-PT-48	2.53	2.54	2.54	2.57
3.0-PT-36	2.60	2.65	2.68	2.64	3.0-PT-48	2.49	2.49	2.49	2.49
3.5-PT-36	2.49	2.51	2.52	2.51	3.5-PT-48	2.34	2.35	2.35	2.35
0.5-R1-36	2.62	2.67	2.68	2.66	0.5-R1-48	2.51	2.53	2.53	2.52
1.0-R1-36	2.50	2.51	2.52	2.51	1.0-R1-48	2.56	2.57	2.57	2.57
1.5-R1-36	2.61	2.63	2.61	2.62	1.5-R1-48	2.57	2.57	2.57	2.57
2.0-R1-36	2.52	2.58	2.58	2.56	2.0-R1-48	2.57	2.56	2.57	2.57
2.5-R1-36	2.57	2.59	2.59	2.58	2.5-R1-48	2.64	2.66	2.66	2.65
3.0-R1-36	2.59	2.63	2.63	2.62	3.0-R1-48	2.35	2.37	2.38	2.37
3.5-R1-36	2.65	2.72	2.73	2.70	3.5-R1-48	2.39	2.41	2.41	2.4
0.5-R2-36	2.68	2.70	2.69	2.69	0.5-R2-48	2.56	2.56	2.56	2.56
1.0-R2-36	2.66	2.68	2.70	2.68	1.0-R2-48	2.54	2.53	2.54	2.54
1.5-R2-36	2.68	2.72	2.72	2.71	1.5-R2-48	2.52	2.52	2.54	2.53
2.0-R2-36	2.69	2.71	2.72	2.71	2.0-R2-48	2.49	2.50	2.50	2.5
2.5-R2-36	2.65	2.67	2.65	2.66	2.5-R2-48	2.69	2.68	2.68	2.68
3.0-R2-36	2.62	2.62	2.62	2.62	3.0-R2-48	2.47	2.48	2.49	2.48
3.5-R2-36	2.59	2.61	2.59	2.59	3.5-R2-48	2.40	2.42	2.43	2.42

Table 5.14 Specific gravity (G) of the fusion products, AAF from TSF

Sample	G_{Avg}	Sample	G_{Avg}	Sample	G_{Avg}
0.2-F1	2.6206	0.2-F2	2.6054	0.2-F3	2.5884
0.4-F1	2.6223	0.4-F2	2.6915	0.4-F3	2.4992
0.6-F1	2.387	0.6-F2	2.3441	0.6-F3	2.3437
0.8-F1	2.2416	0.8-F2	2.2373	0.8-F3	2.3058
1.0-F1	2.2906	1.0-F2	2.2893	1.0-F3	2.3643
1.2-F1	2.221	1.2-F2	2.2759	1.2-F3	2.3613
1.4-F1	2.3327	1.4-F2	2.2966	1.4-F3	2.3753

ppm divided by the molecular weight of Na) in the calculation table, V is the total volume of these solutions (in ml) and m is the mass of the dried sample (in g). Calculations for the computation of the *CEC* are presented in Tables 5.25, 5.26, 5.27, 5.28 and 5.29.

Table 5.15 Specific surface area of the activated fly ash from TSA for 12 h

Sample	W_{AAF} (g)	W_1 (g)	W_2 (g)	W_a (g)	SSA (m^2/g)
0.5-PT-12	1	21.2817	21.2845	0.0028	9.8
1.0-PT-12	1	21.1284	21.1291	0.0007	2.4
1.5-PT-12	1	19.0476	19.0496	0.002	7.0
2.0-PT-12	1	20.2977	20.2997	0.002	7.0
2.5-PT-12	1	20.177	20.1799	0.0029	10.1
3.0-PT-12	1	19.5758	19.5805	0.0047	16.3
3.5-PT-12	1	21.0867	21.0921	0.0054	18.8
0.5-R1-12	1	21.1961	21.2117	0.0156	54.2
1.0-R1-12	1	21.0213	21.0386	0.0173	60.5
1.5-R1-12	1	19.0205	19.042	0.0215	74.8
2.0-R1-12	1	20.2396	20.2407	0.0011	3.8
2.5-R1-12	1	20.1658	20.1763	0.0105	36.7
3.0-R1-12	1	19.5581	19.5649	0.0068	23.6
3.5-R1-12	1	21.0673	21.0738	0.0065	22.6
0.5-R2-12	1	21.1927	21.2165	0.0238	82.6
1.0-R2-12	1	21.0453	21.0772	0.0319	111.4
1.5-R2-12	1	19.0142	19.0429	0.0287	99.8
2.0-R2-12	1	20.2495	20.2516	0.0021	7.3
2.5-R2-12	1	20.1761	20.1911	0.015	52.1
3.0-R2-12	1	19.5693	19.5828	0.0135	46.8
3.5-R2-12	1	21.0779	21.0838	0.0059	20.5

Table 5.16 Specific surface area of the activated fly ash from TSA for 24 h

Sample	W_{AAF} (g)	W_1 (g)	W_2 (g)	W_a (g)	SSA (m^2/g)
0.5-PT-24	1	16.5557	16.5653	0.0096	33.6
1.0-PT-24	1	18.8563	18.8683	0.012	42.0
1.5-PT-24	1	19.5557	19.5654	0.0097	33.9
2.0-PT-24	1	15.495	15.5088	0.0138	48.3
2.5-PT-24	1	19.3984	19.414	0.0156	54.5
3.0-PT-24	1	16.3659	16.3895	0.0236	82.5
3.5-PT-24	1	19.5557	19.5891	0.0334	116.8
0.5-R1-24	1	19.5205	19.5419	0.0214	74.8
1.0-R1-24	1	16.2343	16.2631	0.0288	100.7
1.5-R1-24	1	18.5465	18.5781	0.0316	110.5
2.0-R1-24	1	16.0195	16.0466	0.0271	94.8
2.5-R1-24	1	18.5065	18.532	0.0255	89.1
3.0-R1-24	1	16.019	16.0472	0.0282	98.6
3.5-R1-24	1	15.4955	15.5341	0.0386	134.9
0.5-R2-24	1	15.4953	15.5311	0.0358	125.1
1.0-R2-24	1	16.02	16.0562	0.0362	126.5
1.5-R2-24	1	18.506	18.5505	0.0445	155.6
2.0-R2-24	1	16.019	16.0638	0.0448	156.6
2.5-R2-24	1	18.456	18.5015	0.0455	159.1
3.0-R2-24	1	16.0192	16.0648	0.0456	159.4
3.5-R2-24	1	18.5461	18.5965	0.0504	176.2

Table 5.17 Specific surface area of the activated fly ash from TSA for 36 h

Samples	W_{AAF} (g)	W_1 (g)	W_2 (g)	W_a (g)	SSA (m^2/g)
0.5-PT-36	1	19.5946	19.6818	0.0872	304.9
1.0-PT-36	1	24.3743	24.4673	0.093	325.2
1.5-PT-36	1	20.4189	20.5101	0.0912	318.9
2.0-PT-36	1	20.2824	20.395	0.1126	393.7
2.5-PT-36	1	20.8458	20.8996	0.0538	188.1
3.0-PT-36	1	19.2023	19.2743	0.072	251.7
3.5-PT-36	1	19.3874	19.4522	0.0648	226.6
0.5-R1-36	1	19.5666	19.5701	0.0035	12.2
1.0-R1-36	1	24.3653	24.3935	0.0282	98.6
1.5-R1-36	1	20.4212	20.454	0.0328	114.7
2.0-R1-36	1	20.2638	20.2946	0.0308	107.7
2.5-R1-36	1	20.823	20.843	0.02	69.9
3.0-R1-36	1	19.1861	19.201	0.0149	52.1
3.5-R1-36	1	19.4071	19.4265	0.0194	67.8
0.5-R2-36	1	19.1117	19.1262	0.0145	50.7
1.0-R2-36	1	23.9019	23.916	0.0141	49.3
1.5-R2-36	1	19.943	19.9526	0.0096	33.6
2.0-R2-36	1	19.7887	19.7921	0.0034	11.9
2.5-R2-36	1	20.3728	20.401	0.0282	98.6
3.0-R2-36	1	18.7097	18.7298	0.0201	70.3
3.5-R2-36	1	18.9334	18.9444	0.011	38.5

Table 5.18 Specific surface area of the activated fly ash from TSA for 48 h

Sample	W_{AAF} (g)	W_1 (g)	W_2 (g)	W_a (g)	SSA (m^2/g)
0.5-PT-48	1	21.0629	21.0656	0.0027	9.4
1.0-PT-48	1	19.1612	19.1642	0.003	10.5
1.5-PT-48	1	19.7281	19.7291	0.001	3.5
2.0-PT-48	1	20.3855	20.3862	0.0007	2.4
2.5-PT-48	1	21.0168	21.0184	0.0016	5.6
3.0-PT-48	1	21.1365	21.1422	0.0057	19.9
3.5-PT-48	1	17.7198	17.7209	0.0011	3.8
0.5-R1-48	1	20.5382	20.5416	0.0034	23.7
1.0-R1-48	1	18.6497	18.654	0.0043	29.7
1.5-R1-48	1	19.2171	19.2202	0.0031	21.6
2.0-R1-48	1	19.8704	19.8745	0.0041	28.4
3.0-R1-48	1	20.4902	20.4966	0.0064	44.3
3.5-R1-48	1	20.6248	20.6304	0.0056	39.0
0.5-R2-48	1	15.0873	15.1027	0.0154	53.7
1.0-R2-48	1	20.3855	20.4053	0.0198	68.7
1.5-R2-48	1	15.0873	15.1027	0.0154	53.8
2.0-R2-48	1	19.1612	19.1818	0.0206	71.9
2.5-R2-48	1	20.2638	20.2942	0.0304	105.8
3.0-R2-48	1	20.4212	20.4555	0.0343	119.9
3.5-R2-48	1	20.4212	20.4522	0.031	107.7

Table 5.19 Calculations for the weight loss of the fusion products obtained from the TSA

Sample	W_i (g)	W_f (g)	$W_f - W_i$ (g)	Loss (%)
0.2-F1	72.6223	61.897	10.7253	43
0.2-F2	102.48	102.29	0.19	2.1
0.2-F3	94.39	94.28	0.11	2.2
0.4-F1	67.0333	54.793	12.2403	49
0.4-F2	45.33	44.97	0.36	4
0.4-F3	37.23	37.05	0.18	3.6
0.6-F1	78.6887	65.6329	13.0558	52.2
0.6-F2	41.4	41.07	0.33	3.7
0.6-F3	33.11	32.91	0.2	4
0.8-F1	79.3972	66	13.3972	53.5
0.8-F2	51.46	51.29	0.17	1.9
0.8-F3	43.38	43.28	0.1	2
1.0-F1	81.573	73	8.573	54.2
1.0-F2	50.85	50.66	0.19	2.1
1.0-F3	41.21	40.05	1.16	23.2
1.2-F1	85.6949	72	13.6949	54.7
1.2-F2	65.38	64.76	0.62	6.9
1.2-F3	56.57	56.3	0.27	5.4
1.4-F1	156.23	141.6215	14.6085	58.4
1.4-F2	52.93	51.85	1.08	12
1.4-F3	43.29	42.9	0.39	7.8

5.3.2.6 Determination of the Mineralogical Composition

Mineralogical composition of the fly ash and its residues, AAF, can be determined by employing X-ray diffraction (XRD) analysis by subjecting them to Cu-Kα radiation (using graphite monochromator), and further interpretation of their mineral peaks with the help of mineral data-pdf files of Joint Committee on Powder Diffraction Standards (JCPDS) [35]. It has been reported that the search options in these files includes a variety of variables such as: (a) number of elements, (b) class of compounds (i.e., either minerals or zeolites), (c) common name of the mineral or the zeolite, if known and (d) d-spacing or value of diffraction angle, 2θ. Accordingly, match of a mineral is confirmed if the first, second and third strong lines (i.e., corresponding to three d-spacings present in the XRD data) associated with the mineral are present in the XRD results. Mostly unknown minerals, present in the fly ash and the AAF, were identified by conducting several trials, by using anticipated composition of the material obtained from the XRF analysis. For the sake of brevity, values of d-spacing (i.e., for strong lines) of some of the common zeolites and their PDF codes are listed in Table 5.30 [20, 36]. The XRD diffraction patterns of the TSA samples are presented in Figs. 5.4a–g, 5.5a–g, 5.6a–g and 5.7a–f.

Table 5.20 Chemical composition of the activated fly ash from TSA for 12 h

Sample	% by weight											
	Al_2O_3	BaO	CaO	Fe_2O_3	K_2O	MgO	MnO_2	Na_2O	P_2O_5	SiO_2	SrO	TiO_2
0.5-PT-12	27.93	0.12	1.62	5.38	0.55	0.38	0.05	4.67	0.05	57.73	0.04	1.48
0.5-R1-12	25.48	0.09	1.63	5.04	0.63	0.32	0.04	5.53	0.06	58.26	0.04	1.51
0.5-R2-12	33.84	0.08	1.64	3.73	0.61	0.36	0.04	5.79	0.05	52.56	0.04	1.28
1.0-PT-12	17.65	0.11	1.50	5.10	0.42	0.24	0.04	4.92	0.03	41.63	0.04	1.42
1.0-R1-12	27.19	0.11	1.75	5.78	0.44	0.39	0.05	9.16	NA	52.14	0.04	1.54
1.0-R2-12	13.58	0.08	1.79	4.87	0.40	0.38	0.04	8.53	0.04	68.88	0.04	1.35
1.5-PT-12	39.88	0.09	1.88	4.63	0.33	0.36	0.05	9.29	0.06	41.23	0.04	1.58
1.5-R1-12	26.43	0.09	1.84	5.89	0.29	0.41	0.05	10.19	0.05	53.17	0.04	1.56
1.5-R2-12	25.14	0.10	1.80	5.78	0.30	0.41	0.05	9.42	0.04	55.46	0.04	1.49
2.0-PT-12	19.10	0.11	1.56	5.56	0.24	0.26	0.05	10.10	NA	36.77	0.04	1.42
2.0-R1-12	26.85	0.09	1.93	5.96	0.25	0.41	0.05	11.05	0.04	51.38	0.04	1.60
2.0-R2-12	26.63	0.09	1.93	5.81	0.25	0.46	0.05	11.10	0.04	52.05	0.04	1.54
2.5-PT-12	25.66	0.09	1.92	4.16	0.26	0.31	0.05	10.93	0.06	38.74	0.04	1.54
2.5-R1-12	24.27	0.09	1.95	5.99	0.22	0.43	0.05	11.63	0.04	48.36	0.04	1.55
2.5-R2-12	25.81	0.09	2.00	5.87	0.25	0.47	0.05	12.62	0.04	48.99	0.04	1.53
3.0-PT-12	25.31	0.08	2.36	4.72	0.20	0.35	0.05	10.26	0.04	38.58	0.05	1.54
3.0-R1-12	22.12	0.07	2.31	6.44	0.22	0.39	0.05	12.77	0.04	42.00	0.05	1.52
3.0-R2-12	28.69	0.09	1.97	4.78	0.20	0.35	0.06	14.75	0.04	38.87	0.04	1.47
3.5-PT-12	23.65	0.09	1.76	0.35	0.15	0.36	0.07	14.70	NA	37.93	0.04	1.50
3.5-R1-12	19.69	0.08	2.14	6.14	0.20	0.29	0.05	13.78	0.03	37.86	0.05	1.45
3.5-R2-12	18.90	0.07	2.18	5.95	0.23	0.29	0.05	14.89	0.04	38.55	0.05	1.39

Table 5.21 Chemical composition of the activated fly ash from TSA for 24 h

Sample	% by weight											
	Al_2O_3	BaO	CaO	Fe_2O_3	K_2O	MgO	MnO_2	Na_2O	P_2O_5	SiO_2	SrO	TiO_2
0.5-R1-24	24.21	0.09	1.64	5.53	0.66	0.31	0.05	6.86	0.04	59.12	0.04	1.46
0.5-R2-24	34.27	0.08	1.60	3.47	0.64	0.30	0.05	6.60	0.05	51.56	0.04	1.35
1.0-PT-24	29.24	0.09	1.81	6.45	0.40	0.43	0.06	8.50	NA	51.49	0.04	1.52
1.0-R1-24	25.10	0.09	1.79	6.36	0.42	0.42	0.06	9.09	0.04	55.15	0.04	1.46
1.0-R2-24	33.30	0.07	1.75	4.56	0.41	0.37	0.05	9.50	0.05	48.57	0.04	1.34
1.5-PT-24	25.57	0.10	1.63	5.51	0.38	0.34	0.05	10.60	NA	48.71	0.04	1.51
1.5-R1-24	26.47	0.09	1.85	5.53	0.27	0.41	0.05	9.77	0.04	53.95	0.04	1.51
1.5-R2-24	34.55	0.09	1.86	3.60	0.30	0.39	0.05	8.57	0.05	49.13	0.04	1.40
2.0-PT-24	43.84	0.08	1.80	5.13	0.27	0.41	0.06	8.27	0.04	38.67	0.04	1.41
2.0-R1-24	33.92	0.08	1.91	5.34	0.30	0.40	0.06	11.53	0.03	44.94	0.04	1.46
2.0-R2-24	31.50	0.08	1.84	4.85	0.29	0.36	0.06	13.82	0.04	45.70	0.04	1.42
2.5-PT-24	27.95	0.11	1.90	6.52	0.21	0.42	0.06	11.90	NA	45.83	0.04	1.59
2.5-R1-24	30.17	0.09	1.95	4.46	0.21	0.36	0.06	11.27	0.03	42.90	0.04	1.49
2.5-R2-24	34.51	0.09	1.81	4.05	0.24	0.36	0.05	11.72	0.04	41.88	0.04	1.35
3.0-PT-24	22.41	0.10	1.72	7.18	0.19	0.31	0.07	14.30	NA	39.60	0.04	1.50
3.0-R1-24	18.95	0.08	1.79	6.51	0.21	0.27	0.06	13.46	0.02	39.12	0.04	1.46
3.0-R2-24	33.85	0.09	1.94	5.08	0.24	0.35	0.06	13.65	0.04	36.74	0.04	1.38
3.5-PT-24	14.30	0.09	1.76	4.10	0.21	0.17	0.05	11.77	0.03	28.12	0.04	1.44
3.5-R1-24	18.08	0.09	1.71	5.70	0.22	0.24	0.05	13.27	0.02	35.99	0.04	1.43
3.5-R2-24	16.16	0.08	1.70	3.50	0.22	0.19	0.05	12.43	0.03	35.48	0.04	1.25
0.5-R1-24	24.21	0.09	1.64	5.53	0.66	0.31	0.05	6.86	0.04	59.12	0.04	1.46

Table 5.22 Chemical composition of the activated fly ash from TSA for 36 h

Sample	% by weight											
	Al_2O_3	BaO	CaO	Fe_2O_3	K_2O	MgO	MnO_2	Na_2O	P_2O_5	SiO_2	SrO	TiO_2
0.5-PT-36	23.93	0.12	1.6	5.01	0.55	0.38	0.05	3.98	0.05	52.05	0.04	1.48
0.5-R1-36	16.98	0.05	20.29	9.59	0.71	0.49	0.05	5.57	0.1	37.25	0.05	1.43
0.5-R2-36	20.6	0.05	20.86	9.71	0.71	0.5	0.05	5.6	0.11	39.74	0.05	1.43
1.0-PT-36	18.76	0.04	19.61	90	0.46	0.45	0.04	5.37	0.08	36.96	0.05	1.43
1.0-R1-36	17.51	0.06	21.92	8.8	0.5	0.51	0.04	7.16	0.08	33.29	0.05	1.5
1.0-R2-36	18.31	0.03	19	7.6	0.45	0.48	0.04	5.94	0.07	43.99	0.05	1.27
1.5-PT-36	18.65	0.04	20.65	7.04	0.34	0.5	0.04	5.19	0.09	33.98	0.05	1.46
1.5-R1-36	19.89	0.06	22.5	10.11	0.37	0.47	0.05	9.8	0.07	32.86	0.05	1.52
1.5-R2-36	18.31	0.04	19.56	8.21	0.3	0.48	0.05	7.63	0.06	41.91	0.05	1.26
2.0-PT-36	19.1	0.11	1.56	5.56	0.24	0.26	0.05	10.1	NA	36.77	0.04	1.42
2.0-R1-36	19.74	0.05	21.49	9.02	0.26	0.5	0.05	8.88	0.05	33.96	0.03	1.4
2.0-R2-36	18.91	0.04	19.67	4.48	0.24	0.42	0.05	9.68	0.05	52.05	0.05	1.29
2.5-PT-36	18.6	0.04	19.52	9.82	0.37	0.37	0.05	8.31	0.05	43.59	0.05	1.36
2.5-R1-36	18.25	0.05	21.63	9.81	0.25	0.44	0.05	9.04	0.04	30.64	0.02	1.36
2.5-R2-36	17.94	0.07	24.45	10.25	0.29	0.59	0.05	7.18	0.04	31.89	0.05	1.55
3.0-PT-36	18.03	0.04	20.39	12.11	0.18	0.43	0.06	9.17	0.04	28.26	0.05	1.39
3.0-R1-36	18.83	0.05	22.37	11.38	0.23	0.46	0.06	10.19	0.05	41.37	0.05	1.39
3.0-R2-36	13.13	0.05	17.94	10.8	0.21	0.32	0.05	13.24	0.05	37.29	0.04	1.17
3.5-PT-36	17.74	0.02	21.87	5.4	0.17	0.29	0.05	13.81	0.04	43.83	0.05	1.31
3.5-R1-36	17.35	0.04	20.01	10.87	0.22	0.33	0.05	12.24	0.04	35.04	0.05	1.29
3.5-R2-36	17.39	0.02	18.94	10.81	0.21	0.4	0.05	9.92	0.05	30.77	0.05	1.09

Table 5.23 Chemical composition of the activated fly ash from TSA for 48 h

Sample	% by weight											
	Al_2O_3	BaO	CaO	Fe_2O_3	K_2O	MgO	MnO_2	Na_2O	P_2O_5	SiO_2	SrO	TiO_2
0.5-PT-48	36.99	0.09	1.69	3.86	0.66	0.29	0.05	6.13	0.05	48.74	0.04	1.45
0.5-R1-48	24.21	0.09	1.64	5.53	0.66	0.31	0.05	6.86	0.04	59.12	0.04	1.46
0.5-R2-48	18.3	0.03	19	7.08	0.62	0.41	0.04	4.4	0.08	43.21	0.05	1.31
1.0-PT-48	13.19	0.05	20.18	10.96	0.48	0.37	0.05	5.1	0.12	51.41	0.05	1.49
1.0-R1-48	18.43	0.41	19.51	10.11	0.33	0.39	0.05	8.18	0.06	37.12	0.05	1.36
1.0-R2-48	18.99	0.05	19.07	1.17	0.49	0.38	0.06	8.36	0.07	45.95	0.04	1.37
1.5-PT-48	19.06	0.04	19.56	8.94	0.42	0.41	0.05	6.06	0.08	47.84	0.05	1.39
1.5-R1-48	26.47	0.09	1.85	5.53	0.27	0.41	0.05	9.77	0.04	53.95	0.04	1.51
1.5-R2-48	18.06	0.03	18.97	9.48	0.26	0.36	0.05	7.55	0.05	38.26	0.05	1.29
2.0-PT-48	17.66	0.04	19.37	8.69	0.28	0.33	0.05	9.65	0.04	41.7	0.05	1.33
2.0-R1-48	19.17	0.04	20.15	9.88	0.24	0.41	0.05	9.56	0.04	34.3	0.05	1.32
2.0-R2-48	15.19	0.04	18.24	11.46	0.21	0.36	0.06	11.87	0.05	42.79	0.04	1.21
2.5-PT-48	18.38	0.04	20.25	5.58	0.26	0.38	0.05	9.09	0.04	39.78	0.05	1.39
2.5-R1-48	11.94	0.06	21.35	10.73	0.25	0.47	0.05	8.15	0.04	41.62	0.05	1.39
2.5-R2-48	10.77	0.27	13.96	8.52	0.17	0.13	0.04	5.55	0.04	48.74	0.04	1.02
3.0-PT-48	22.41	0.1	1.72	7.18	0.19	0.31	0.07	14.3	NA	39.6	0.04	1.5
3.0-R1-48	14.41	0.06	23.11	9.6	0.27	0.58	0.05	13.43	0.05	25.26	0.05	1.44
3.0-R2-48	17.28	0.05	20.11	8.77	0.22	0.38	0.05	13.4	0.05	37.27	0.05	1.21
3.5-PT-48	14.3	0.09	1.76	4.1	0.21	0.17	0.05	11.77	0.03	28.12	0.04	1.44
3.5-R1-48	18.08	0.09	1.71	5.7	0.22	0.24	0.05	13.27	0.02	35.99	0.04	1.43
3.5-R2-48	12.3	0.03	17.54	9.8	0.19	0.34	0.05	12.49	0.05	35.8	0.04	1.03
Zeolite 4A	46.04	0.03	0.22	0	0	0	0.02	16.87	0.02	36.57	0.01	0.24
Hopper Fly ash#	26.05	0.14	1.88	5.14	0.66	0.39	0.16	0.054	10.5	63.85	0.25	1.52

Table 5.24 Chemical composition of the activated fly ash from TSA (by fusion)

Sample	% by weight										
	Al_2O_3	BaO	CaO	Fe_2O_3	K_2O	MgO	MnO_2	Na_2O	P_2O_5	SiO_2	SrO	TiO_2
RFA	26.05	0.197	1.88	5.14	0.66	0.39	0.06	0.054	0.16	63.85	0.04	1.52
0.2-F1	26.06	0.09	1.765	4.53	0.567	0.256	0.052	8.716	0.145	56.193	0.037	1.269
0.2-F2	24.55	0.09	2.036	4.866	0.67	0.318	0.057	10.442	0.171	52.863	0.044	1.472
0.2-F3	26.96	0.06	1.88	4.66	0.621	0.24	0.056	8.157	0.15	33.174	0.044	1.401
0.4-F1	24.45	0.08	1.79	4.57	0.54	0.26	0.055	16.94	0.158	47.51	0.041	1.268
0.4-F2	25.77	0.07	1.815	4.627	0.543	0.236	0.058	17.342	0.164	45.32	0.044	1.293
0.4-F3	25.94	0.05	1.76	4.85	0.559	0.267	0.06	17.874	0.17	45.06	0.042	1.276
0.6-F1	22.9	0.04	1.568	4.648	0.557	0.221	0.055	24.424	0.161	43.418	0.038	1.137
0.6-F2	23.61	0.05	1.69	4.808	0.611	0.24	0.059	24.016	0.166	40.06	0.041	1.334
0.6-F3	24.39	0.05	1.604	4.193	0.589	0.232	0.053	24.45	0.164	39.838	0.04	1.181
0.8-F1	21.28	0.04	1.55	4.166	0.673	0.26	0.067	31.32	0.162	33.287	0.04	1.197
0.8-F2	21.01	0.04	1.48	4.027	0.616	0.225	0.067	33.216	0.151	34.583	0.039	1.141
0.8-F3	19.06	0.06	1.512	4.58	0.618	0.159	0.077	31.443	0.142	26.645	0.046	1.249
1.0-F1	20.57	0.04	1.35	3.769	0.59	0.123	0.054	31.122	0.131	30.77	0.04	1.131
1.0-F2	20.67	0.06	1.408	3.79	0.55	0.138	0.052	32.27	0.136	32.6	0.039	1.138
1.0-F3	18.57	0.07	1.41	3.793	0.55	0.131	0.054	33.592	0.131	26.958	0.04	1.157
1.2-F1	19.74	0.10	1.293	3.778	1.502	0.09	0.054	33.603	0.117	25.872	0.04	1.098
1.2-F2	19.34	0.06	1.26	3.638	0.51	0.078	0.052	34.592	0.112	27.917	0.037	1.069
1.2-F3	18.93	0.07	1.24	3.444	0.597	0.077	0.051	35.42	0.114	29.527	0.038	1.06
1.4-F1	17.22	0.08	1.137	3.096	0.398	0.131	0.047	42.416	0.107	28.425	0.033	0.976
1.4-F2	17.40	0.06	1.105	3.085	0.390	0.128	0.042	41.385	0.109	28.005	0.034	0.975
1.4-F3	17.62	0.04	1.134	3.091	0.434	0.099	0.049	40.659	0.115	27.868	0.035	0.989

Table 5.25 Calculation table for the cation exchange capacity of the activated fly ash obtained from TSA, hydrothermally for 12 h

AAF	Concentration	Mol. Wt.	Epm	Extracted Vol.	Sample wt.	CEC
	C_{Na}	m_{Na} (g)	(mg/l)	V (ml)	m (g)	meq./100 g
0.5-PT-12	875	23	38.043	27	1	103
1.0-PT-12	525	23	22.826	27	1	62
1.5-PT-12	980	23	42.6086	27	1	115
2.0-PT-12	930	23	40.434	27	1	109
2.5-PT-12	4335	23	188.478	27	1	509
3.0-PT-12	2655	23	115.434	27	1	312
3.5-PT-12	2420	23	105.217	27	1	284
0.5-R1-12	1575	23	68.478	27	1	185
1.0-R1-12	2792	23	121.391	27	1	328
1.5-R1-12	2795	23	121.521	27	1	328
2.0-R1-12	2421	23	105.26	27	1	284
2.5-R1-12	2631	23	114.391	27	1	309
3.0-R1-12	3180	23	138.26	27	1	373
3.5-R1-12	2436	23	105.913	27	1	286
0.5-R2-12	1701	23	73.956	27	1	200
1.0-R2-12	1967.5	23	85.543	27	1	231
1.5-R2-12	1949	23	84.739	27	1	229
2.0-R2-12	2119.5	23	92.152	27	1	249
2.5-R2-12	1720	23	74.782	27	1	202
3.0-R2-12	1585.5	23	68.934	27	1	186
3.5-R2-12	1697	23	73.956	27	1	198

5.3.2.7 Determination of Crystallinity

The percentage crystallinity of zeolites was computed by using XRD diffraction pattern of the sample. The important parameter referred is the relative intensity of the mineral peaks of the sample and that of a standard sodium zeolite 4A. The crystallinity of the AAF of each step of the TSA, was calculated from the ratio of sum of minimum ten numbers of higher relative intensities of the zeolitic product with that of the intensities of the standard zeolite, as expressed in Eq. 5.6 [18, 37]:

$$\%\ \text{Crystallinity} = \sum_{i=1}^{10} PIP_i / \sum_{i=1}^{10} PIRZ_i \tag{5.6}$$

where, PIP_i = ith number of higher peak intensity of the products and $PIRZ_i$ = ith number of higher peak intensity of the reference zeolite.

The results of the crystallinity of the samples are listed in Table 5.31.

Table 5.26 Calculation table for the cation exchange capacity of AAF from TSA for 24 h

AAF	Concentration	Mol. Wt.	Epm	Extracted Vol.	Sample wt.	*CEC*
	C_{Na}	m_{Na} (g)	(mg/l)	V (ml)	m (g)	meq./100 g
0.5-PT-24	2442	23	106.174	27	1	287
1.0-PT-24	2377	23	103.348	27	1	279
1.5-PT-24	2523	23	109.695	27	1	296
2.0-PT-24	2345	23	101.956	27	1	275
2.5-PT-24	1128	23	49.043	27	1	132
3.0-PT-24	2950	23	128.261	27	1	346
3.5-PT-24	1544	23	67.13	27	1	181
0.5-R1-24	1600	23	69.565	27	1	188
1.0-R1-24	1650	23	71.739	27	1	194
1.5-R1-24	4007.5	23	174.239	27	1	470
2.0-R1-24	2010	23	87.391	27	1	236
2.5-R1-24	5293	23	230.13	27	1	621
3.0-R1-24	3395	23	147.608	27	1	399
3.5-R1-24	2010	23	87.391	27	1	236
0.5-R2-24	1580	23	68.695	27	1	185
1.0-R2-24	1550	23	67.391	27	1	182
1.5-R2-24	7184	23	312.348	27	1	843
2.0-R2-24	4257.5	23	185.108	27	1	500
2.5-R2-24	3824	23	166.261	27	1	449
3.0-R2-24	4080.5	23	177.413	27	1	479
3.5-R2-24	1725	23	75	27	1	203

5.3.3 Determination of Crystallite Size

The crystallite size of a mineral, from its X-ray diffractogram, has been evaluated by applying Scherrer's relationship. However, assigning values and units of variables (viz., peak width and diffraction angle) in this relationship is contentious and such computations are quite cumbersome and time consuming. Hence, development of a 'look up chart' using d-spacing, peak width and crystallite size of a mineral would be quite prudent, for materials like fly ash and its residues obtained from the alkali activation.

5.3.3.1 Modified Computation of Crystallite Size

With due reference to the previous researchers much attention was given to clarify selection and utilization of appropriate units of the main variables [38–41], used in Scherrer's formula, presented in Eq. 5.7.

$$CS = \mathrm{K} \cdot \lambda/(\beta \cdot \cos \theta) \tag{5.7}$$

Table 5.27 Calculation table for the cation exchange capacity of AAF from TSA for 36 h

AAF	Concentration C_{Na}	Mol. Wt. m_{Na} (g)	Epm (mg/l)	Extracted Vol. V (ml)	Sample wt. m (g)	*CEC* meq./100 g
0.5-PT-36	1418	23	61.652	27	1	167
1.0-PT-36	1808	23	78.608	27	1	212
1.5-PT-36	1670	23	72.608	27	1	196
2.0-PT-36	1280	23	55.652	27	1	150
2.5-PT-36	640	23	27.826	27	1	75
3.0-PT-36	671	23	29.173	27	1	79
3.5-PT-36	3900	23	169.565	27	1	458
0.5-R1-36	2170	23	94.347	27	1	255
1.0-R1-36	1720	23	74.782	27	1	202
1.5-R1-36	1970	23	85.652	27	1	231
2.0-R1-36	1675	23	72.826	27	1	197
2.5-R1-36	720	23	31.304	27	1	85
3.0-R1-36	495	23	21.521	27	1	58
3.5-R1-36	2570	23	111.739	27	1	302
0.5-R2-36	1040	23	45.217	27	1	122
1.0-R2-36	4795	23	208.478	27	1	563
1.5-R2-36	2240	23	97.391	27	1	263
2.0-R2-36	1550	23	67.391	27	1	182
2.5-R2-36	1490	23	64.782	27	1	175
3.0-R2-36	1280	23	55.652	27	1	150
3.5-R2-36	1015	23	44.130	27	1	119

Accordingly, β (i.e., full width at half maximum in 2θ°) adopted from the X-ray diffractograms was converted to radian, while cos θ, used in Eq. 5.7 was calculated by using the value of θ in radian. Also, λ(=1.540 Å), was converted to nm unit (i.e., λ = 0.154 nm), before substituting it in Eq. 5.7. Here K is equal to 0.9 (the average shape factor for a spherical crystallite).

The conventional way of getting a correct value of *CS* is a cumbersome task and it might also yield manual errors, as well, particularly if proper units of the variables are not input in Eq. 5.7. With this in view and for the precise and quick determination of *CS*, as listed in Table 5.32, look up chart has been developed, as depicted in Fig. 5.8. Each chart exhibits a regular trend of variation between the *CS* and d-spacing (corresponding to the 2θ°) of various minerals present in the samples, for different β (ranging from 0.0502 to 1.0706 in 2θ° unit) values. Thus getting the value of *CS* from Fig. 5.8 looks very simple, except for an interpolation in case of an intermediate value of β, between adjacent plots.

Highlights of the Look up Chart

Graphical relationship, as depicted in Fig. 5.8, was developed between crystallite size and d-spacing for a given FWHM. Figure 5.8 elucidates interdependency

Table 5.28 Calculation table for the cation exchange capacity of AAF from TSA for 48 h

AAF	Concentration C_{Na}	Mol. Wt. m_{Na} (g)	Epm (mg/l)	Extracted Vol. V (ml)	Sample wt. m_{AFA} (g)	*CEC* meq./100 g
0.5-PT-48	950.5	23	41.326	27	1	112
1.0-PT-48	1092	23	47.478	27	1	128
1.5-PT-48	1951.5	23	84.848	27	1	229
2.0-PT-48	873	23	37.956	27	1	102
2.5-PT-48	1524	23	66.261	27	1	179
3.0-PT-48	1359	23	59.087	27	1	160
3.5-PT-48	1326	23	57.652	27	1	156
0.5-R1-48	1200.5	23	52.196	27	1	141
1.0-R1-48	1451.5	23	63.108	27	1	170
1.5-R1-48	1371.5	23	59.630	27	1	161
2.0-R1-48	3631	23	157.869	27	1	426
2.5-R1-48	1578	23	68.608	27	1	185
3.0-R1-48	1520.5	23	66.108	27	1	178
3.5-R1-48	1166.5	23	50.717	27	1	137
0.5-R2-48	1211.5	23	52.674	27	1	142
1.0-R2-48	1284	23	55.826	27	1	151
1.5-R2-48	2037.5	23	88.587	27	1	239
2.0-R2-48	2090	23	90.869	27	1	245
2.5-R2-48	1664	23	72.348	27	1	195
3.0-R2-48	1277	23	55.522	27	1	150
3.5-R2-48	954	23	41.478	27	1	112

between the *CS* and d-spacing, which is indirectly proportional to each other. This is in full agreement with Eq. 5.7, for a constant value of β (the FWHM). Further, it is observed that the *CS* varies significantly for even a small change in β (refer Table 5.32). From most of the plots depicted in Fig. 4.4, major revelation is that *CS* is almost unchanged corresponding to higher values of d-spacing beyond 5 Å.

Such a graphical trend contradicts the trend that *CS* should decrease with a decreasing 2θ° or an increasing d-spacing (refer Eq. 5.7). However, this establishes a quantifiable trend of variation between the *CS* and d-spacing, which has not been clearly defined by previous researchers [38]. A comparison of the value of the *CS* from the 'look up chart' vis-à-vis the value obtained from Eq. 5.7, was carried out and results are presented in Table 5.33. It is quite apparent that the value of *CS* is larger, when it is computed by using Eq. 5.7, in radian unit. On the other hand, the value of *CS* is quite small when β is an input in degree units. Incidentally, it is quite interesting to note that this observation matches with the *CS* obtained from the 'look up chart'. Further, based on a re-analysis of Fig. 5.8, it is inferred that the crystallite size falls in a specific range for each value of β, as listed in Table 5.34. Each range of *CS* is found to have a range of d-spacing, as well. In combination with Fig. 5.8, a fact, which is emphasized by Tables 5.32 and 5.34 is that value of d-spacing on

Table 5.29 Calculation table for the cation exchange capacity of fusion products, AAF obtained from TSA, fusion technique

AAF	Concentration C_{Na}	Mol. Wt. m_{Na} (g)	Epm (mg/l)	Extracted Vol. V (ml)	Sample wt. m_{AFA} (g)	*CEC* meq./100 g
0.2-F1	1026.6	23	44.634	27	1	120.5
0.2-F2	834.2	23	36.269	27	1	97.9
0.2-F3	972.1	23	42.265	27	1	114.1
0.4-F1	1583.6	23	68.852	27	1	185.9
0.4-F2	1941.6	23	84.417	27	1	227.9
0.4-F3	2055.6	23	89.373	27	1	241.3
0.6-F1	2522.2	23	109.66	27	1	296.1
0.6-F2	2333	23	101.434	27	1	273.9
0.6-F3	1505.4	23	65.452	27	1	176.7
0.8-F1	2413.9	23	104.952	27	1	283.4
0.8-F2	2327.6	23	101.2	27	1	273.2
0.8-F3	2024.7	23	88.030	27	1	237.7
1.0-F1	2327.6	23	101.2	27	1	273.2
1.0-F2	2236	23	97.2173913	27	1	262.5
1.0-F3	2926.8	23	127.2521739	27	1	343.6
1.2-F1	2897.4	23	125.973913	27	1	340.1
1.2-F2	2855.3	23	124.1434783	27	1	335.2
1.2-F3	1711.4	23	74.40869565	27	1	200.9
1.4-F1	5014.5	23	218.0217391	27	1	588.7
1.4-F2	2489.8	23	108.2521739	27	1	292.3
1.4-F3	2879.1	23	125.1782609	27	1	338.0

higher side of the range corresponds to a constant value of *CS*. Thus, this table can also be referred for getting a preliminary idea about the crystallite size corresponding to a known value of β and d-spacing, from the XRD pattern of a material.

5.3.3.2 FT-IR Study

The FTIR analysis of the samples was done by employing following the standard methodology. Approximately, 15 mg of the sample and 1 g of KBr were mixed, milled and ground in an agar mortar and pestle for 5 min to make a fine and smooth powder consisting of the uniform sized particles. A quarter of the prepared mixture ($\sim$0.25 g) was pressed with a steel die at a pressure of 10 ton/cm^2 to prepare a pellet. The IR spectra of the sample were recorded by employing a Perkin Elmer make, Paragon 1000 PC FT-IR spectrometer, in the range 4000–400 cm^{-1} [33]. In order to interpret IR spectrums of common zeolitic minerals, the frequency range corresponding to different modes of structural vibrations was employed [7, 18, 33, 37, 42–44]. The FT-IR spectrum of the zeolitized residues is presented in Fig. 5.9a–c.

Table 5.30 Tools for mineral identification by employing JCPDS data file (1998)

Common minerals and zeolites (legends used in the XRD)	d-spacing (Å) of strong lines in descending order of peak intensity	PDF number
Quartz (Q)	3.34, 4.25, 2.45, 2.28, 2.23, 2.12, 1.81	83-2465
Mullite (M)	5.39, 3.39, 2.69, 2.54, 2.20,3.42, 1.50	73-1253
Zeolite P2	7.05, 7.13, 3.19, 4.10, 3.12, 5.04, 4.93	80-0700
Sodalite (S)	3.62, 6.27, 2.56, 2.03, 2.80, 1.52, 1.57	81-0704
Zeolite A	12.04, 8.51, 6.95, 5.38, 3.63, 4.01, 2.92	74-1185
Zeolite X (X)	14.46, 8.86, 7.55, 5.74, 2.89, 3.34, 6.26	83-0915
Zeolite Y	14.30, 5.67, 3.77, 2.85, 4.37, 2.20, 8.73	38-0239
Hydroxysodalite	3.63, 6.28, 2.56, 2.09, 2.81, 1.73, 4.44	11-0401
Herschelite	2.93, 4.32, 9.36, 5.03, 2.89, 3.88, 6.89	19-1178
KM zeolite/	3.19, 3.25, 2.97, 5.07, 8.35, 7.15, 5.38	30-0902
Philipsite	3.19,7.10, 4.10, 5.04, 3.47, 2.68, 2.75	46-1427
Cancrinite (C)	4.68, 6.34, 3.24, 3.66, 2.11, 2.42, 2.74	75-2318
Na-P1 (P)	3.17, 7.10, 4.10, 2.68, 5.02, 1.97, 2.36	39-0219
Linde F	3.00, 3.07, 3.08, 6.96, 2.81, 3.47, 2.96	25-0619
Herschelite	3.67, 2.59, 6.36, 2.11, 1.76, 2.73 1.58	31-1271
Faujasite (F)	14.3, 5.71, 8.75, 7.51, 4.39, 3.78, 2.86	12-0228
Hydroxycancrinite	3.26, 4.71, 3.68, 2.76, 2.44, 4.18, 2.12	28-1036
Tobermorite	3.08, 11.3, 2.98, 2.82, 1.94, 5.48, 3.53	19-1364
Analcime (A)	3.43, 5.60, 2.92, 2.22, 4.85, 1.74, 2.69	19-1180
K-Chabazite	2.93, 4.32, 1.72, 9.50, 3.95, 2.59, 2.29	12-0194

5.3.3.3 Determination of Morphological Characteristics

The morphological characteristics of the zeolitic materials were studied by employing field emission gun-scanning electron microscopy (FEG-SEM) model No. JSM-7600F (resolution X25 to 1,000,000) and micrographs were obtained to investigate the shape, size and surface texture of crystals, as depicted in Fig. 5.10a–t for the hydrothermal and fusion product, respectively. In addition, these micrographs were further investigated, by employing energy dispersive spectrometry, EDS, for determining number of elements present on the surface. The results of the EDS analysis are tagged with their corresponding FEG-SEM images, as exhibited in Fig. 5.10a–t.

5.3.3.4 Interaction of AAF in Acidic Medium

Acid soluble components of the zeolitic material were identified by monitoring the interaction of highly zeolitized product of the TSA with a strong acid. In this context, 1.2 N HCl (35 %, $pH = 0.7$), a strong acid, was used in this study. In order to quantify solubility of the residue, AAF, in such an acidic solution (i.e., similar to

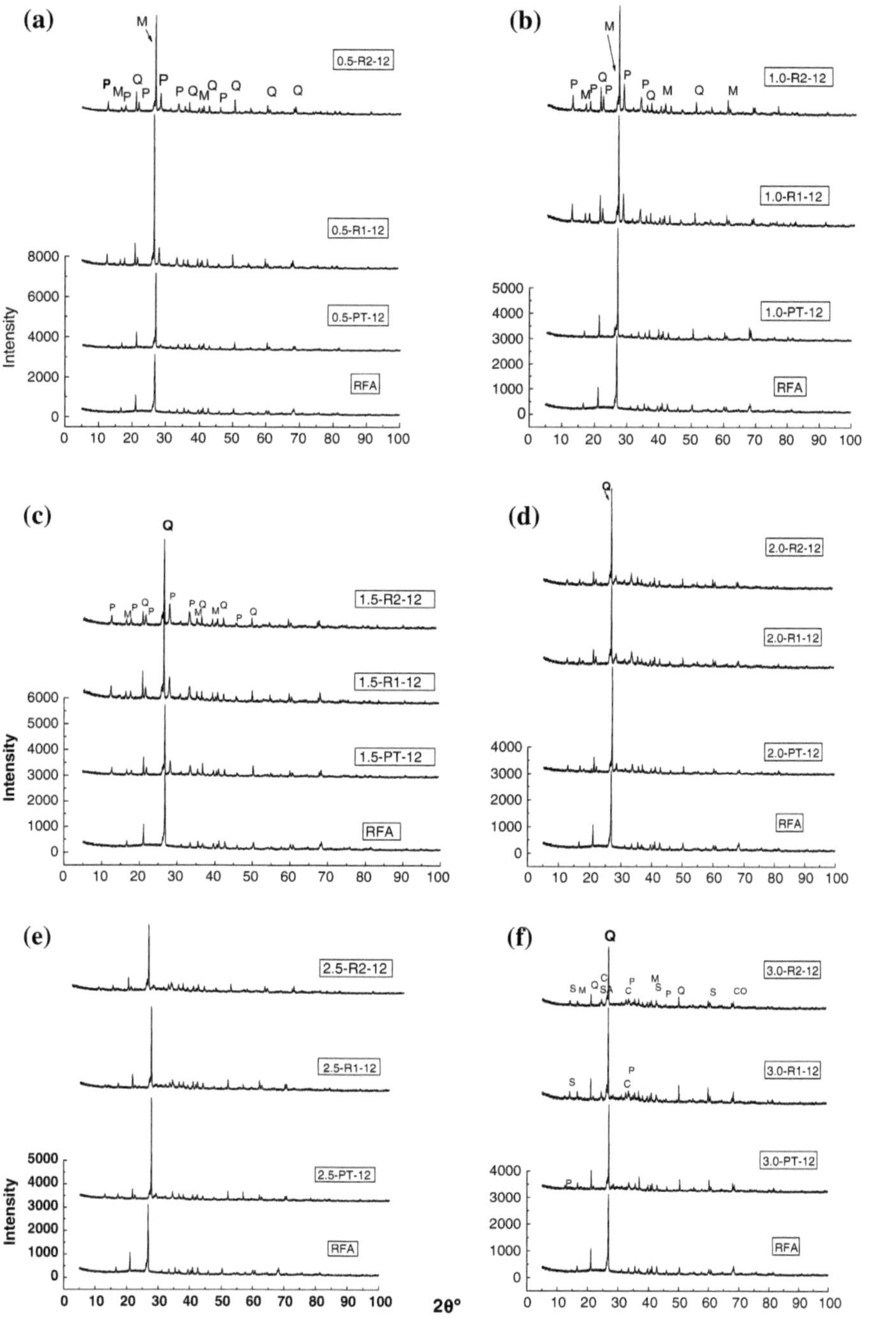

Fig. 5.4 **a**, **b** XRD patterns of the AAFs of the TSA obtained hydrothermally (12 h and **a** 0.5 M, **b** 1.0 M NaOH). **c**–**f** XRD patterns of the AAFs of the TSA obtained hydrothermally (12 h and **c** 1.5 M, **d** 2.0 M, **e** 2.5 M, **f** 3.0 M NaOH). **g** XRD patterns of the AAFs of the TSA obtained hydrothermally (12 h and **g** 3.5 M NaOH)

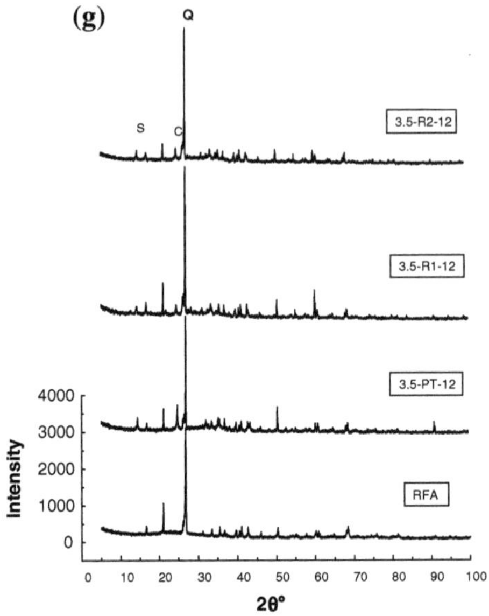

Fig. 5.4 (continued)

an acid mine drainage and sewage waste), the dose of AAF for interaction with the acid was widely varied over a range (e.g., 18, 30, 50, and 150 mg per 30 ml of the acidic solution, corresponding to 0.6, 1.0, 1.66 and 5.0 g/l, respectively). The interaction period was varied from 6 to 24 h (at an increment of 6 h) for each dose of the AAF. Furthermore, the concentration of the soluble ions (viz., Si, Al, Fe, Na, Ca, K, Mg), which leached out from the AAF to the HCl solution, was determined with the help of an ICP-AES. The result of the analysis is presented in Table 5.35 and Figs. 5.11 and 5.12.

5.3.3.5 Thermal Analysis of AAF

Thermogravimetric analysis (TGA) and differential thermal analysis (DTA) set up (Diamond TG/DTA, PERKIN ELMER, USA) were employed for thermal analysis of some significantly zeolitized residues of the alkali activated fly ash (AAF). Around 8–10 mg of the sample, contained in a α-Al_2O_3 crucible, was heated, continuously, up to 1000 °C, at a rate of 10 °C/min. During the analysis, a controlled environment (i.e. an inert gas, N_2, flow rate between 0 and 1000 ml/min, under vacuum 10^{-2} Torr) was maintained for monitoring actual transitions (viz., mass change, dehydration, oxidation, mineral phase, crystallization and all endothermic and exothermic activities) undergone by the samples. The results of these analyses are presented in graphical form in Figs. 5.13 and 5.14. In general, loss in weight from a TGA curve indicates presence of water and volatile matters in

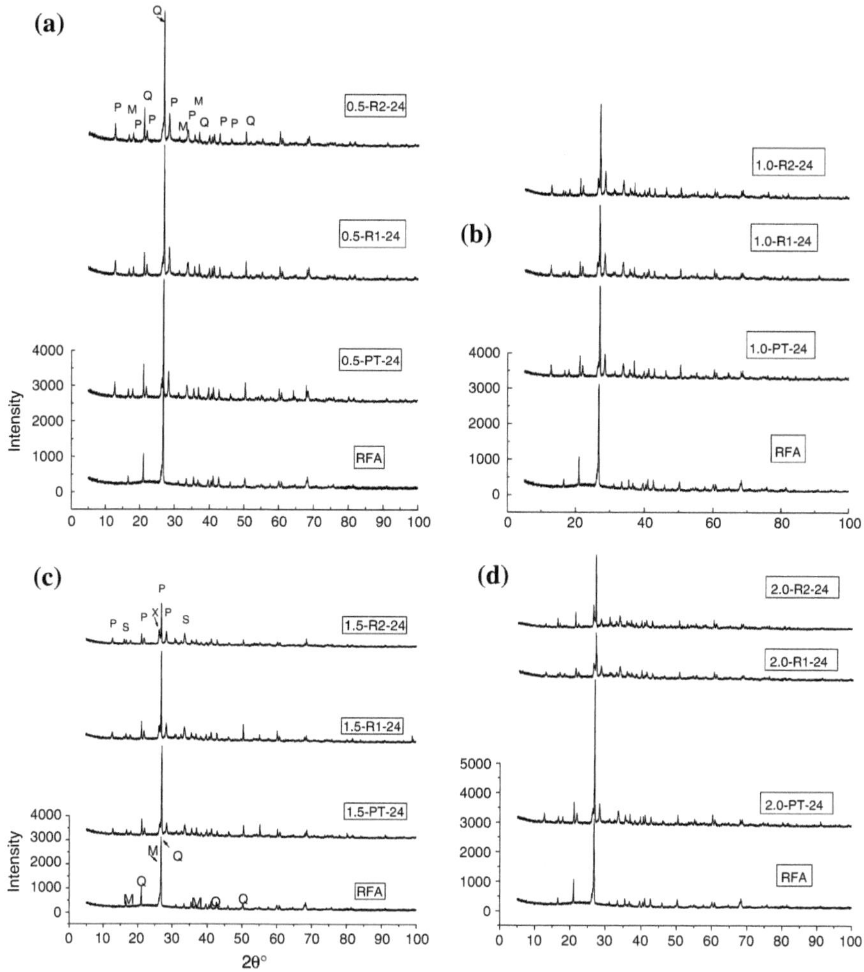

Fig. 5.5 **a**, **b** XRD patterns of the AAFs of the TSA (24 h and **a** 0.5 M, **b** 1.0 M NaOH). **c**–**f** XRD patterns of the AAFs of the TSA (24 h and **c** 1.5 M, **d** 2.0 M, **e** 2.5 M, **f** 3.0 M NaOH). **g** XRD patterns of the AAFs of the TSA (24 h and **g** 2.5 M NaOH)

the samples. In the DTA curve, the peaks are revelation to endothermic and exothermic reactions corresponding to its sagging (peak down) and hogging (peak up) trends, respectively, as depicted in Figs. 5.13 and 5.14 for superior hydrothermal and fusion products, respectively.

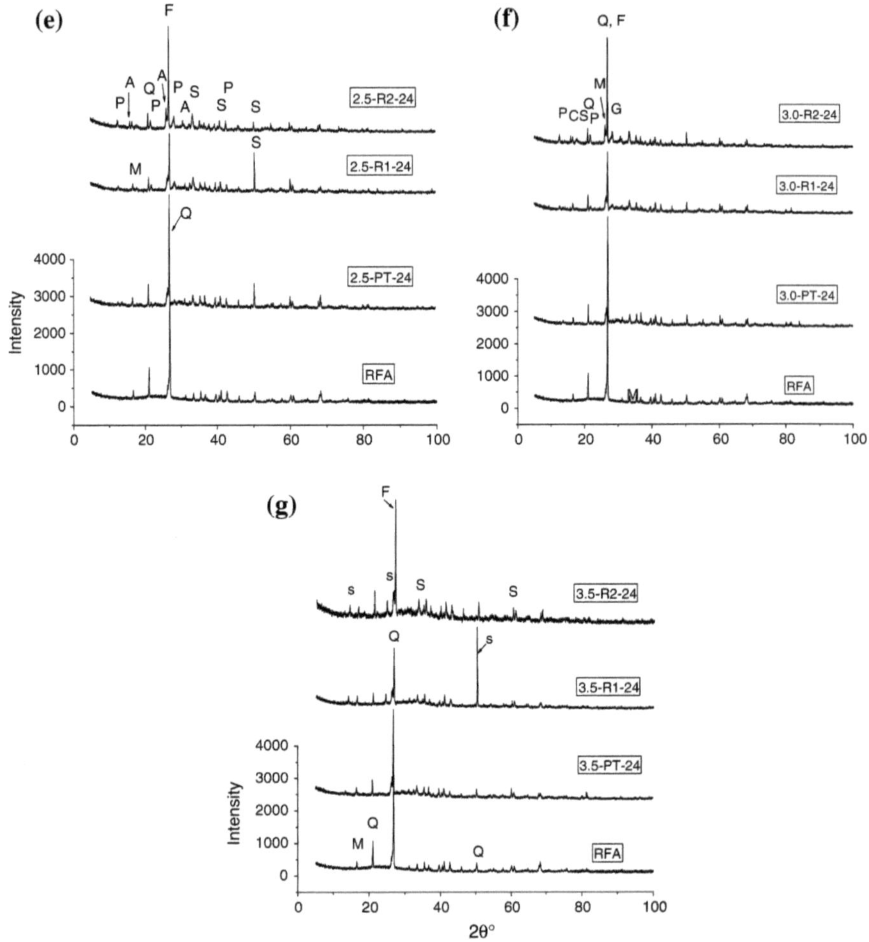

Fig. 5.5 (continued)

5.3.3.6 NMR Study of AAF

The nuclei type (Si or Al) present in some significantly zeolitized residues of the AAF were investigated with the help of a nuclear magnetic resonance (NMR) spectrometer set up (Mercury Plus 300 MHz, VARIAN, USA). The spectrometer consists of a 5 mm auto switchable probe with PFG (1H/13C/31P/19F), 5 mm Dual Broad Band probe with PFG for multinuclear NMR (^{13}C, ^{15}N, ^{27}Al, ^{31}P, ^{29}Si, ^{77}Se, ^{119}Sn, ^{125}Te, ^{199}Hg, ^{51}V, ^{7}Li etc.).

In the present work, ^{27}Al NMR and ^{29}Si NMR were carried out to evaluate the transition in these elements in and out of their tetrahedra, which are understood to get polymerized to form different types of zeolites. In fact, under influence of amagnetic field, nuclei of Si and Almay get energized and spun, differently. This

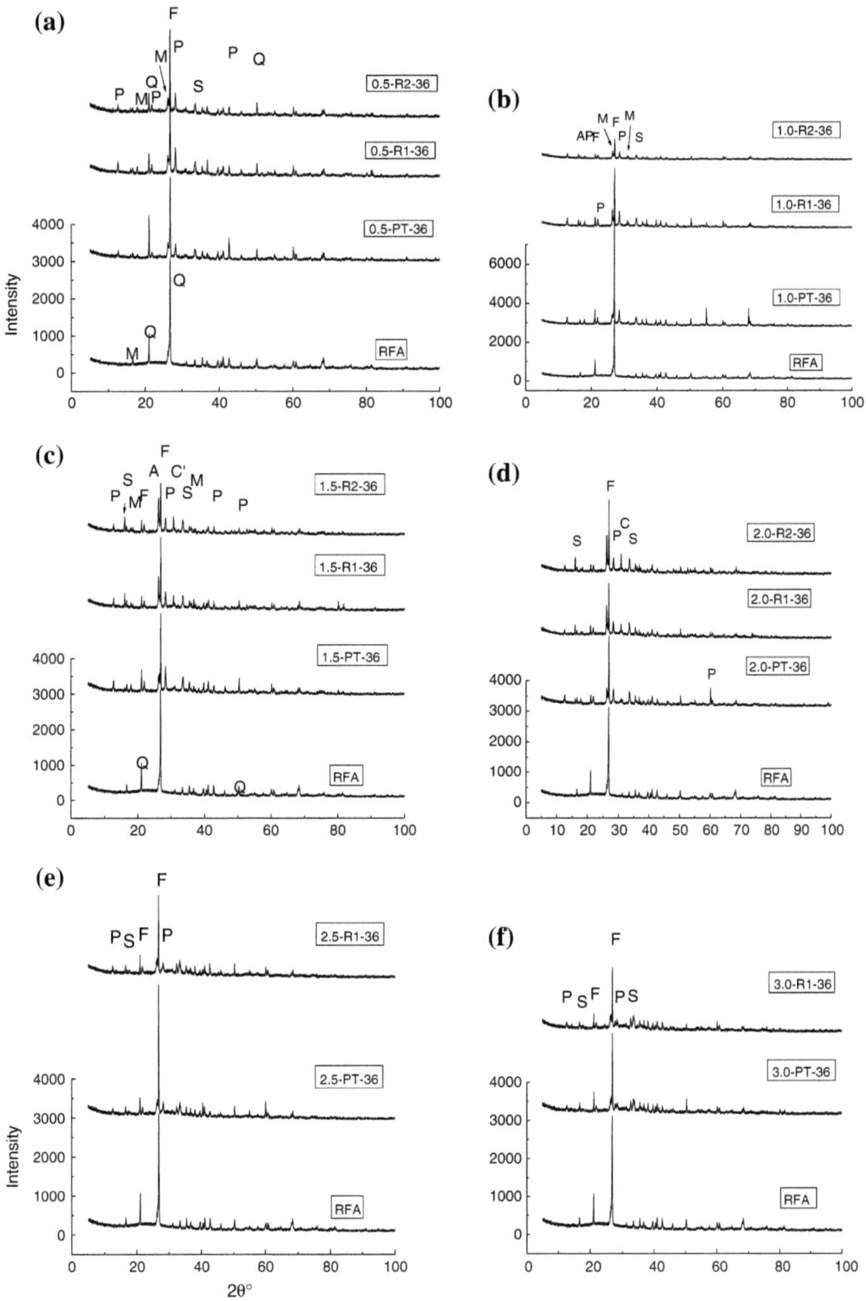

Fig. 5.6 XRD patterns of the AAFs of the TSA (36 h and **a** 0.5 M and **b** 1.0 M NaOH). **c–f** XRD patterns of the AAFs of the TSA (36 h and **c** 1.5 M, **d** 2.0 M, **e** 2.5 M, **f** 3.0 M NaOH). **g** XRD patterns of the AAFs of the TSA (36 h and **g** 3.5 M NaOH)

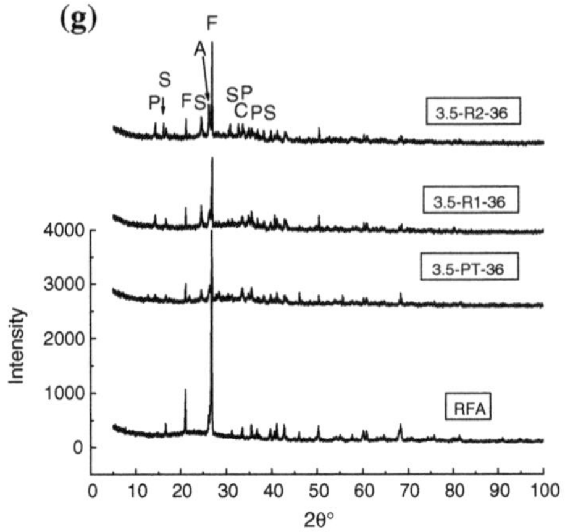

Fig. 5.6 (continued)

transition can be interlinked to a type of the nucleus and its chemical environment. Such energized nucleus is allowed to respond freely and finally come back to an equilibrium position, which conforms to a peak in the NMR spectra. Accordingly, the spectra (i.e. graphical relation between absorbance and spin energy or frequency, in Hz, of a nucleus, as compared to that, in MHz, of spectrometer, normally represented in ppm) of AAF samples are presented in Figs. 5.15, 5.16 and 5.17.

5.4 Analysis of the Results of Hopper Ash and Lagoon Ash

Based on analysis of the results of the experimentation presented above, with the hopper ash and the lagoon ash, both have undergone wide variations in specific gravity, silicon aluminium ratio, chemical composition, cation exchange capacity, mineralogy and morphological characteristics after conventional hydrothermal alkali activation. This could be observed by monitoring the results of the characterization tools (viz., X-ray fluorescence, X-ray diffractograms and micrographs) for the two ashes and their products, after conventional hydrothermal treatment. Moreover, the analysis of the results of the three step activation of the superior ash is to come up in the next chapter.

5.4.1 *Superiority of Hopper Ash Over Lagoon Ash*

In fact, fly ash disposal at thermal power plants is either through wet- or dry-disposal systems. As these disposal practices result in different characteristics of fly ashes [45], their interaction with alkali and hence, their zeolitization characteristics

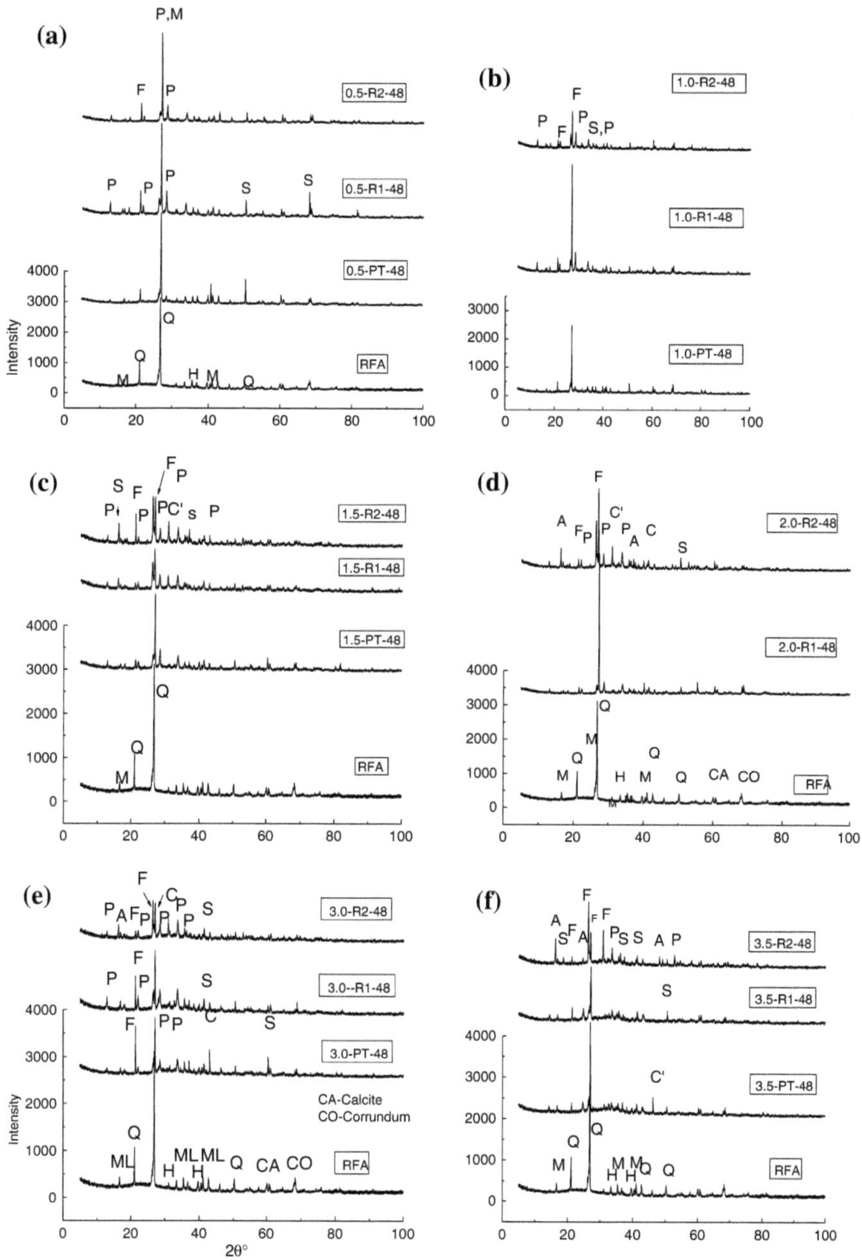

Fig. 5.7 **a**, **b** XRD patterns of the AAFs of the TSA (48 h and **a** 0.5 M, **b** 1.0 M NaOH). **c**–**f** XRD patterns of the AAFs of the TSA (48 h and **c** 1.5 M, **d** 2.0 M, **e** 3.0 M, **f** 3.5 M NaOH)

Table 5.31 Calculated values of the crystallinity of TSA products obtained hydrothermally

Steps of activation	Activation period (h)	Molarity of NaOH (M)						
		0.5	1.0	1.5	2.0	2.5	3.0	3.5
PT	12	29.21	28.25	44.05	26.49	27.00	39.73	35.41
R1		28.12	38.24	40.73	33.97	30.85	30.89	28.69
R2		32.63	37.39	34.14	32.02	31.54	31.98	29.26
PT	24	36.99	38.12	34.98	30.94	34.87	32.70	33.08
R1		33.93	42.24	39.66	48.28	50.13	39.76	52.74
R2		33.15	37.80	60.60	36.78	32.65	33.83	38.48
PT	36	42.01	26.68	42.49	39.11	28.56	35.06	39.31
R1		39.42	40.90	43.34	48.66	37.96	41.96	47.59
R2		35.45	45.80	54.86	41.82	29.00	33.00	40.70
PT	48	36.03	31.94	36.54	31.73	50.65	61.14	37.33
R1		44.03	53.03	54.51	50.56	56.00	60.42	50.27
R2		32.48	51.29	70.73	55.00	65.00	76.18	53.18

may also get affected. In such a scenario, extensive investigations were conducted to characterize the physical, chemical, mineralogical and morphological characteristics of fly ashes from both the systems, and their residues after hydrothermal treatment. Accordingly, it has been demonstrated that the hydrothermal treatment of the fly ash collected from the hopper yields much better zeolite as compared to its counterpart from the lagoon [9].

Field emission gun-scanning electron microscopy (FEG-SEM) and X-ray diffraction (XRD) analyses of these residues confirms that the hopper ash attains exceptionally higher cation exchange capacity and SiO_2/Al_2O_3 ratios (*SAR*), which correspond to superior fly ash zeolites as compared to its counterpart. In other words, the superiority of the original hopper ash (OHA) over the original lagoon ash (OLA) for synthesizing fly ash zeolites gets ascertained. The parametric variation from OHA and OLA to their activated counterparts (viz., activated hopper ash, AHA and activated lagoon ash, ALA) have been described in the following sections [9].

5.4.1.1 Variation of Specific Gravity

From Fig. 5.18 it is clear that the specific gravity, *G*, of the original hopper ash, OHA, (=2.18) is almost same as that of the original lagoon ash, OLA, (=2.13).

Though, in general, *G* values for the ALA and AHA samples are seen to increase with an increase in the molarity of the NaOH ($M \leq 2.0$), randomness in the value of *G* for the AHA with time, *T* is noticeable. Incidentally, the residues AHA7 (i.e., the hopper ash treated with 1.0 M NaOH solution for 36 h) and ALA6 (i.e., the lagoon ash treated with 1.0 M NaOH solution for 24 h), exhibit very high *G* values (2.73 and 2.53, respectively). The increase in *G* value can be attributed to the

Table 5.32 Sample calculations for the *CS* for various peaks present in the XRD pattern of the fly ash sample

Angle (2θ°)	FWHM (2θ°)	2θ°/2 (rad)	*β* (rad)	λ (nm)	cos θ	*CS* (nm)	d-spacing (Å)
16.63	0.2007	0.1451	0.00350	0.154	0.9895	40	5.32
21.04	0.2007	0.1835	0.00350	0.154	0.9832	40	4.22
26.82	0.2007	0.2339	0.00350	0.154	0.9728	41	3.32
31.17	0.2007	0.2719	0.00350	0.154	0.9633	41	2.86
33.45	0.2007	0.2918	0.00350	0.154	0.9577	41	2.67
35.45	0.2007	0.3092	0.00350	0.154	0.9526	42	2.53
36.7	0.2676	0.3201	0.00467	0.154	0.9492	31	2.44
39.66	0.2342	0.3459	0.00409	0.154	0.9408	36	2.27
40.5	0.2007	0.3533	0.00350	0.154	0.9383	42	2.22
41.05	0.2007	0.3580	0.00350	0.154	0.9366	42	2.19
42.67	0.2676	0.3722	0.00467	0.154	0.9315	31	2.11
45.97	0.2007	0.4010	0.00350	0.154	0.9207	43	1.97
48.35	0.4015	0.4217	0.00700	0.154	0.9124	21	1.88
50.3	0.2342	0.4387	0.00409	0.154	0.9053	37	1.81
54.19	0.4015	0.4727	0.00700	0.154	0.8904	22	1.69
54.99	0.2676	0.4796	0.00467	0.154	0.8872	33	1.66
57.77	0.2007	0.5039	0.00350	0.154	0.8757	45	1.59
60.14	0.2007	0.5246	0.00350	0.154	0.8655	45	1.53
60.81	0.2007	0.5304	0.00350	0.154	0.8626	45	1.52
64.74	0.2007	0.5647	0.00350	0.154	0.8448	47	1.43
68.36	0.3346	0.5963	0.00584	0.154	0.8274	29	1.37
75.83	0.2007	0.6614	0.00350	0.154	0.7891	50	1.25
80.06	0.4015	0.6983	0.00700	0.154	0.7659	26	1.19
81.39	0.4015	0.7099	0.00700	0.154	0.7584	26	1.18
83.93	0.8029	0.7321	0.01401	0.154	0.7438	13	1.15
91.02	0.9792	0.7939	0.01708	0.154	0.7011	11	1.07

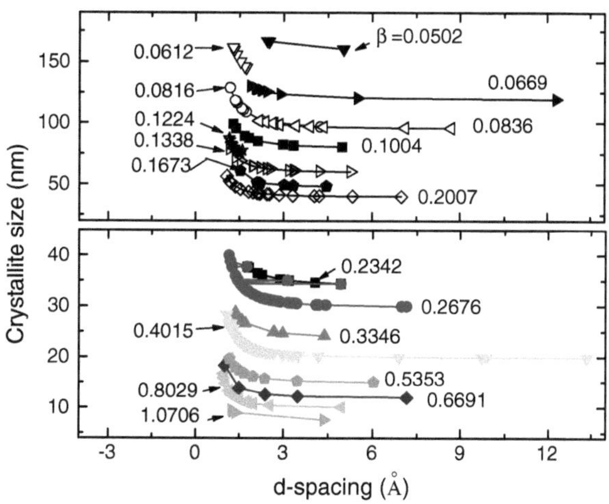

Fig. 5.8 Depiction of d-spacing and crystallite size as "*look up charts*" for different values of FWHM from 0.0502 to 1.0706

Table 5.33 Comparison of values of *CS* obtained from Scherrer's formula and look up chart

Sample	Angle	CS (nm) from Eq. (4.7)		
		Units for β and θ		Look up chart
	(2θ°)	Degree	Radian	
RFA	26.67	22	81	81
	26.95	6	20	20
	27.09	6	20	20
	26.79	12	49	49
	27.06	6	20	20
	27.04	9	31	31
	26.82	10	41	41
AAF	27.09	17	61	61

Table 5.34 Probable ranges of crystallite size and d-spacing for a value of FWHM

FWHM (2θ°)	*CS* (nm)	d-spacing (Å)	Sample
0.0502	160–167	5.02–2.43	AAF
0.0612	145–162	1.73–1.27	AAF
0.0669	119–130	5.54–1.93	AAF
0.0836	95–130	5.60–1.21	AAF
0.1004	80–100	5.02–1.27	AAF
0.1224	75–88	1.51–1.11	AAF
0.1338	60–79	4.34–1.23	AAF
0.1673	48–61	4.56–1.51	AAF
0.2007	40–57	3.37–1.07	RFA, AAF
0.2342	34–40	5.01–1.13	RFA, AAF
0.2676	30–40	4.31–1.07	RFA, AAF
0.3346	24–29	4.37–1.39	RFA, AAF
0.4015	20–28	4.10–1.13	RFA, AAF
0.5353	15–20	3.66–1.20	AAF
0.6691	12–18	3.46–1.20	RFA
0.8029	10–20	2.56–0.94	RFA, AAF
1.0706	7–9	4.31–1.26	RFA

etching of the ash particles, which helps in the escape of entrapped air and gases from the inner regions of the fly ash particle, due to alkali treatment [2, 46]. On the other hand, treatment of the fly ash with high molar solutions (i.e., $M \geq 2.0$), results in a significant increment (up to 2.67) and reduction (up to 2.35) in the *G* values of the ALA and AHA samples, respectively. This could be attributed to a gradual increase in the dissolution of the lighter ingredients (e.g., cenospheres) present in the lagoon ash, as compared to the hopper ash, for which the reaction is faster, even at lower molarities. Also, as depicted in micrographs (refer Fig. 5.24), the AHA samples develop more internal pores, which justify the increased zeolite formation and results in lower *G* values. The increased zeolite formation in AHA

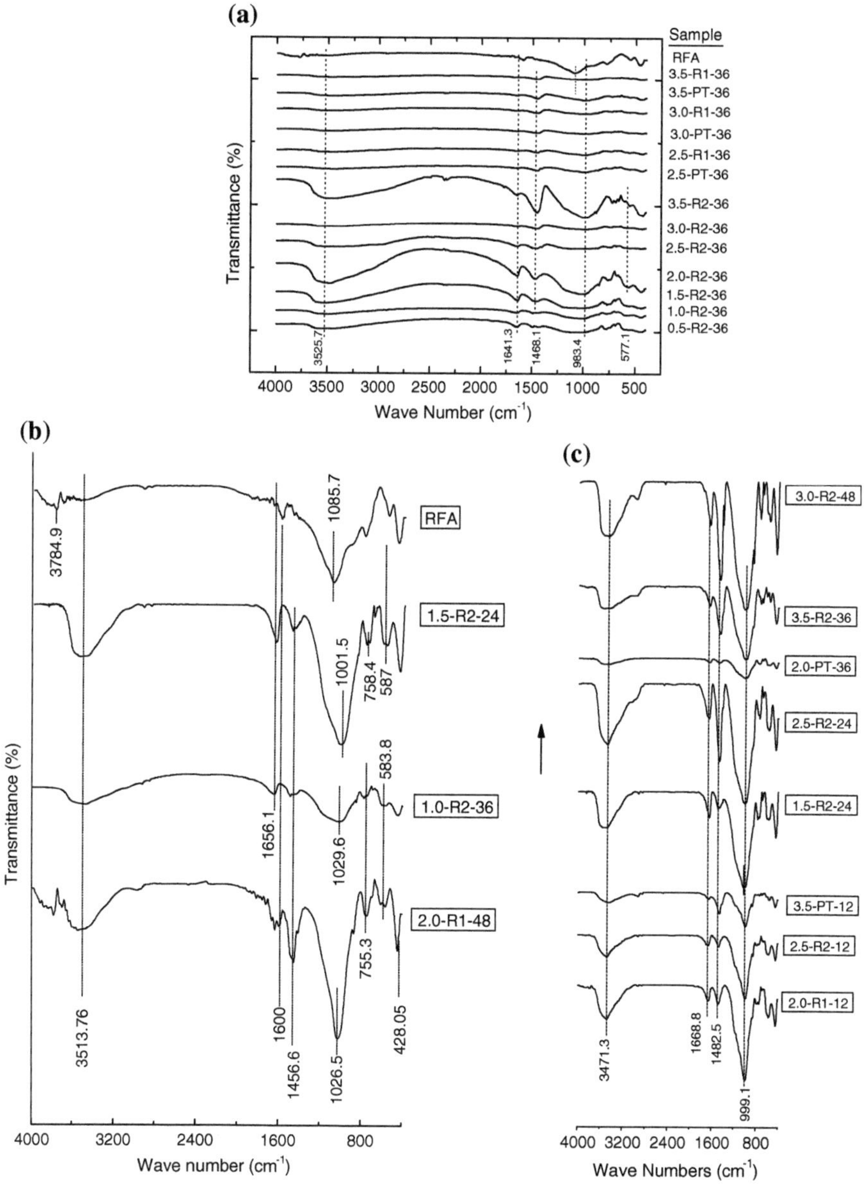

Fig. 5.9 **a** FT-IR transmittance spectra of the superior residues of the TSA obtained from hydrothermal technique. **b** FT-IR transmittance spectra (magnified) of the superior residues of the TSA obtained from hydrothermal technique. **c** FT-IR transmittance spectra of the superior residues of the TSA obtained from hydrothermal technique

(a)

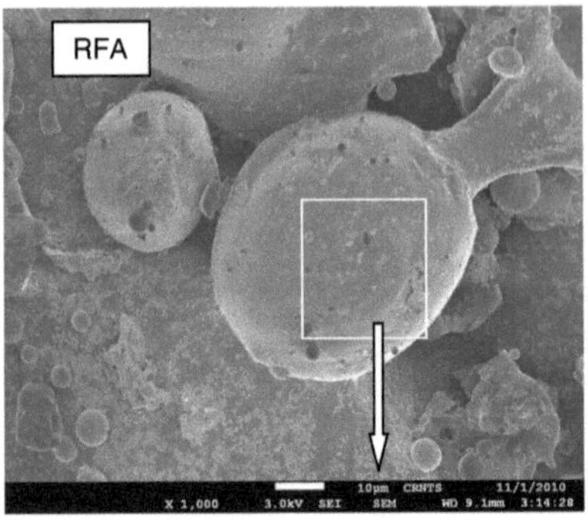

Element	Weight (%)	Atomic (%)
O	50.5	67.65
Al	14.62	11.61
Si	23.24	17.74
K	0.49	0.27
Ca	1.55	0.83
Ti	1.65	0.74
Fe	1.08	0.41
Cu	0	0
Zn	0	0
Pt	6.88	0.76

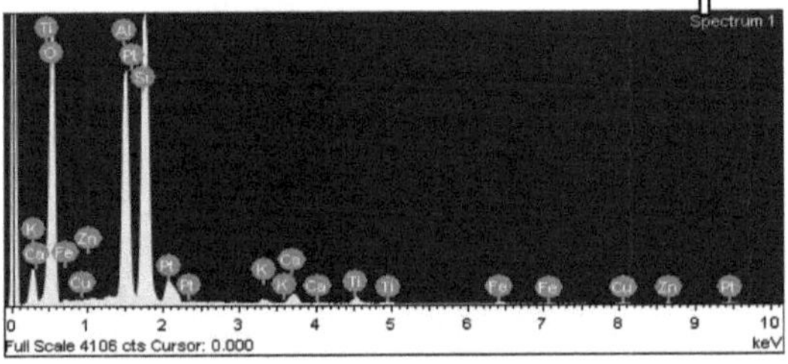

(b)

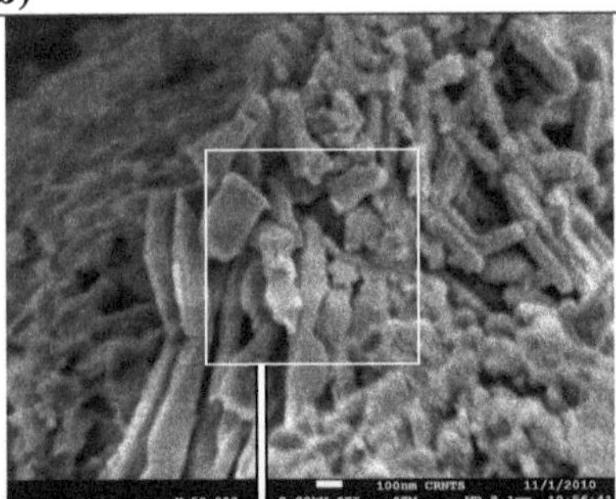

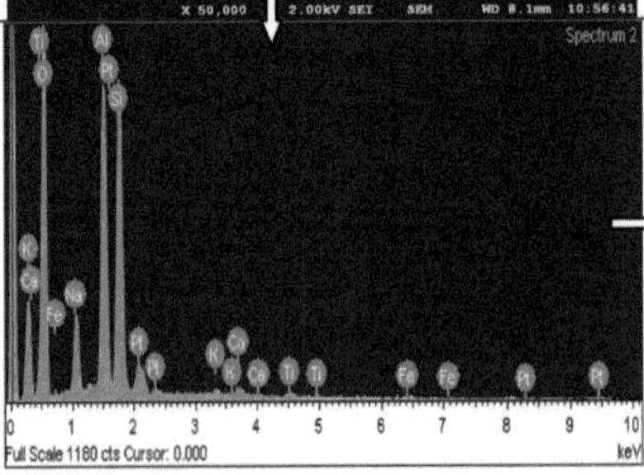

Element	Weight (%)	Atomic (%)
O	68.37	78.73
Na	5.04	4.04
Al	11.60	7.92
Si	12.89	8.46
K	0.43	0.20
Ca	0.62	0.29
Ti	0.41	0.16
Fe	0.65	0.21

◀ **Fig. 5.10** **a** FEG-SEM micrographs and EDS data of the raw fly ash. **b** FEG-SEM micrographs and EDS data of the activated fly ash (0.5-PT-12). **c** FEG-SEM micrographs and EDS data of the activated fly ash (0.5-R1-12). **d** FEG-SEM micrographs and EDS data of the activated fly ash (1.5-PT-12). **e** FEG-SEM micrographs and EDS of the activated fly ash (1.5-R1-12). **f** FEG-SEM micrographs and EDS data of the activated fly ash (1.5-R2-12). **g** FEG-SEM micrographs and EDS data of the activated fly ash (0.5-PT-24). **h** FEG-SEM micrographs and EDS data of the activated fly ash (1.0-R1-24). **i** FEG-SEM micrographs and EDS data of the activated fly ash (1.5-R2-24). **j** FEG-SEM micrographs and EDS data of the activated fly ash (1.0-R1-12). **k** FEG-SEM micrographs and EDS data of the activated fly ash (2.0-R1-12). **l** FEG-SEM micrographs and EDS data of the activated fly ash (2.5-R2-12). **m** FEG-SEM micrographs and EDS data of the activated fly ash (3.0-PT-12). **n** FEG-SEM micrographs and EDS data of the activated fly ash (3.5-PT-12). **o**-*i* FEG-SEM micrographs and EDS spectrum of the residue1.5-R2-24. **o**-*ii* EDS results 1–4 of the residue, 1.5-R2-24. **o**-*iii* FEG-SEM micrographs of the activated fly ash ***a*** 1.5-R2-24 at different field and ***b*** 3.0-R2-24. **p** FEG-SEM micrographs of the activated fly ash by TSA *i* 2.5-R1-36 *ii* 3.5-R2-36 *iii* 1.5-R2-48 and *iv* 3.5-R1-48. **q** FEG-SEM micrographs of TSA, fusion products, 1.4-F1 at different resolutions *i*, *ii* 1500 and *iii* 10000. **r** FEG-SEM micrographs of TSA, fusion products, 1.0-F1 at different resolutions *i*, *ii* 10000 *iii* 30000. **s**-*i* FEG-SEM micrographs of TSA, fusion products, 1.0-F3 at different resolutions ***a*** 20000, ***b*** 50000, and ***c*** 100000. **s**-*ii* FEG-SEM micrographs of TSA, fusion products, 0.4-F3 at 30000 resolution. **t** FEG-SEM micrographs and EDS data of the standard zeolite 4A

samples is also justified by the significant dissolution of SiO_2 (e.g., from 63.85 %, for the OHA, to 36.96 %, for AHA7, which is 27 %) and Al_2O_3 (from 26.05 %, to 18.76 %, which is 7 %). This dissolution of SiO_2 and Al_2O_3 also results in enhanced pores in the residues (refer micrographs in Fig. 5.24).

5.4.2 *Variation of Silicon Aluminium Ratio* (SAR)

From Table 5.36 it can be observed that though both OHA and OLA belong to class F fly ash (ASTM C 618-08), as $SiO_2 + Al_2O_3 + Fe_2O_3$ is more than 90 %, the percentage of SiO_2 in the OHA is higher by 5 %, as compared to the OLA. This excess amount of SiO_2 in the OHA may be in the amorphous aluminosilicate glass phase, which is responsible for higher silica dissolution, as shown in Table 5.37. From the data presented in this table, it is clear that the *SAR* of all the AHA samples is in the range 1.79–1.97, which matches well with the *SAR* value for the zeolites Na-P1 and Analcime [47]. Such a reduced *SAR* (lower than 2.45 in the OHA) can be attributed to either a loss of SiO_2 or a gain of Al_2O_3 in these residues. This also indicates mineralogical and morphological transitions, which further confirm zeolitization of the OHA.

In contrast, Table 5.37 also exhibits an increase in SiO_2 in some of the residues (i.e., ALA2 to ALA6) of the hopper ash, as compared to the lagoon ash. Also, Al_2O_3 has been noticed to be on the higher side, as compared to the OLA, in these samples except for ALA1.

However, such an increase in the oxides in the residues of OLA is unexpected and could not be substantiated, as neither SiO_2 nor Al_2O_3 were added externally.

(c)

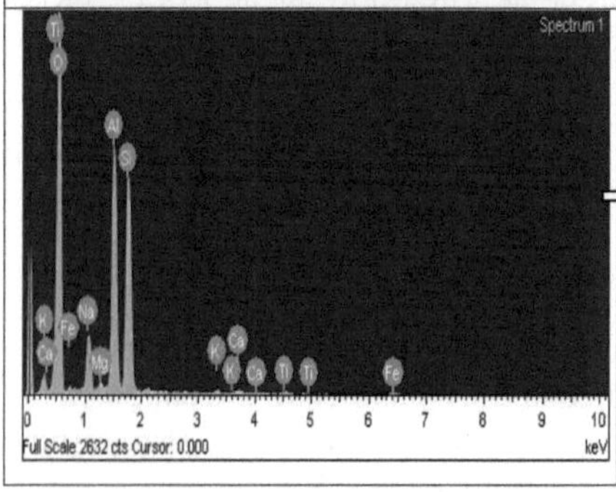

Element	Weight (%)	Atomic (%)
O	56.59	69.25
Na	4.45	3.79
Mg	0.48	0.39
Al	17.13	12.43
Si	18.76	13.07
K	0.41	0.21
Ca	0.48	0.23
Ti	0.53	0.22
Fe	1.18	0.41

(d)

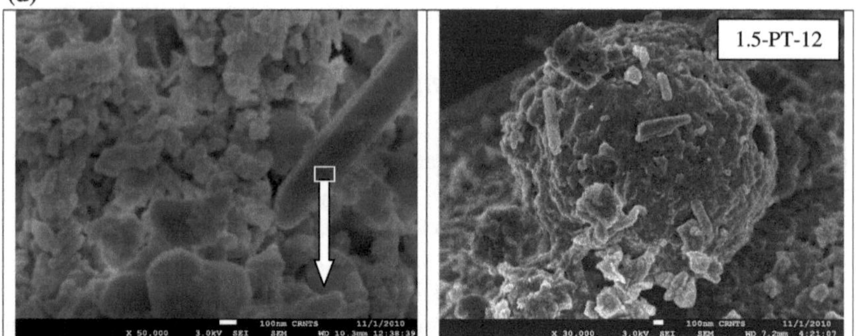

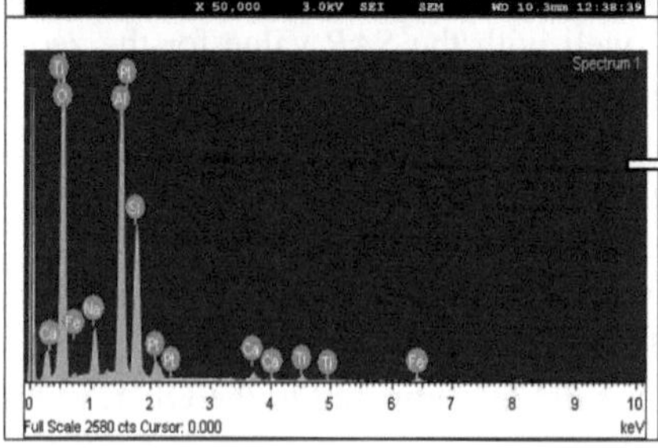

Element	Weight (%)	Atomic (%)
O	70.91	80.76
Na	4.15	3.29
Al	14.14	9.55
Si	8.59	5.57
Ca	0.59	0.27
Ti	0.63	0.24
Fe	0.99	0.32

Fig. 5.10 (continued)

(e)

Element	Weight (%)	Atomic (%)
O	26.27	54.07
Na	2.27	3.25
Al	12.17	14.84
Si	12.64	14.82
Ca	1.88	1.55
Ti	1.40	0.96
Fe	7.60	4.48
Pt	35.77	6.04
Bi	0.00	0.00

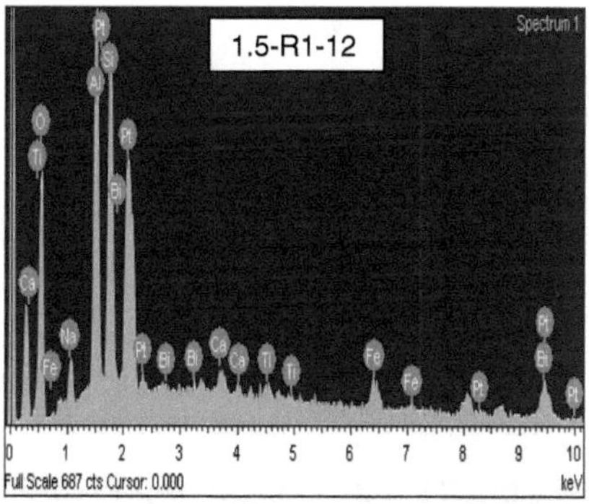

Element	Weight (%)	Atomic (%)
C	17.48	32.17
O	29.82	41.19
Na	2.77	2.66
Al	7.81	6.40
Si	12.45	9.80
Ca	2.35	1.30
Ti	1.18	0.55
Fe	10.53	4.17
Cu	0.00	0.00
Re	1.40	0.17
Pt	14.19	1.61

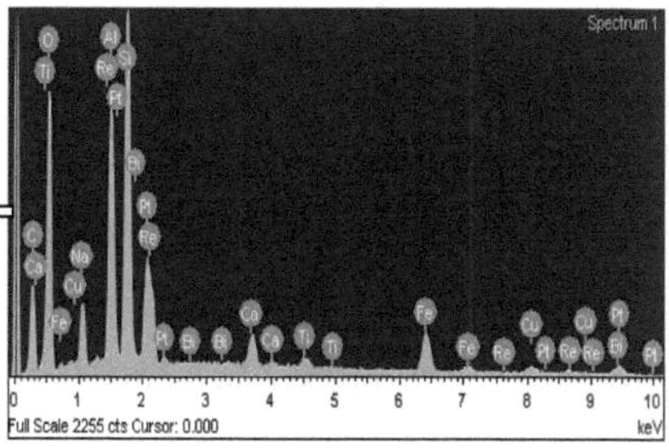

Fig. 5.10 (continued)

Hence, the present generation researchers focus should be on an extensive future study to explore, the chemistry involved in such process.

Furthermore, it can be inferred that the increase in Al_2O_3 and decrease in SiO_2 in the ALA samples contributes to a significant lowering of their *SAR* up to 1.26 (refer Table 5.37). Also, a decrease in the *SAR* of the ALA samples with an increase in the

(f)

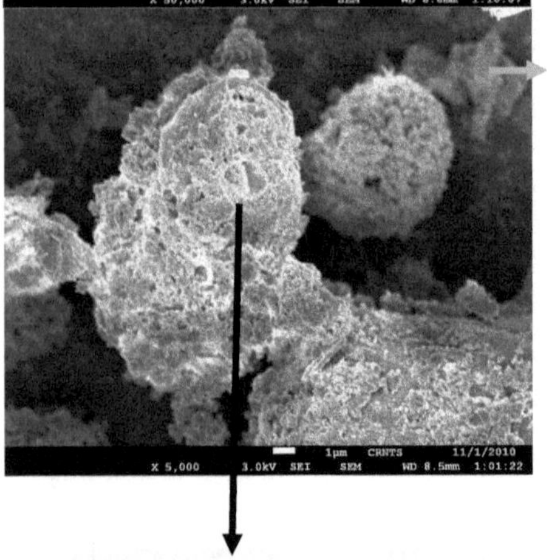

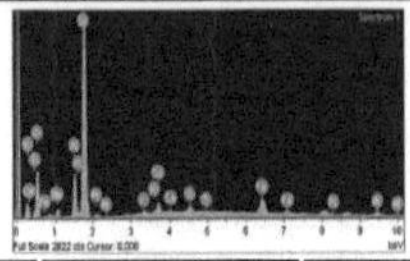

Element	Weight (%)	Atomic (%)
O	42.77	60.79
Na	2.00	1.98
Al	5.32	4.49
Si	29.05	23.52
K	0.70	0.41
Ca	2.56	1.45
Ti	2.89	1.37
Fe	14.70	5.99

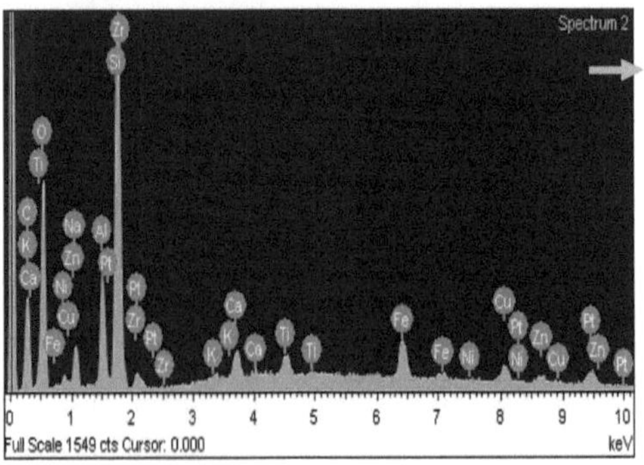

Element	Weight (%)	Atomic (%)
C	30.84	42.36
O	44.07	45.45
Na	1.92	1.38
Al	2.70	1.65
Si	9.65	5.67
K	0.27	0.11
Ca	1.59	0.65
Ti	1.66	0.57
Fe	7.31	2.16

Fig. 5.10 (continued)

molarity, *M* and time, *T* of the treatment may have some effects on their increasing *G*, as shown in Fig. 5.18. Thus, based on the *SAR* values, it can be concluded that the lagoon ash used in this study is a suitable material for the synthesis of a group of low-silica zeolites (2 > *SAR* > 1.26), such as Na-P1 and Hydroxysodalite.

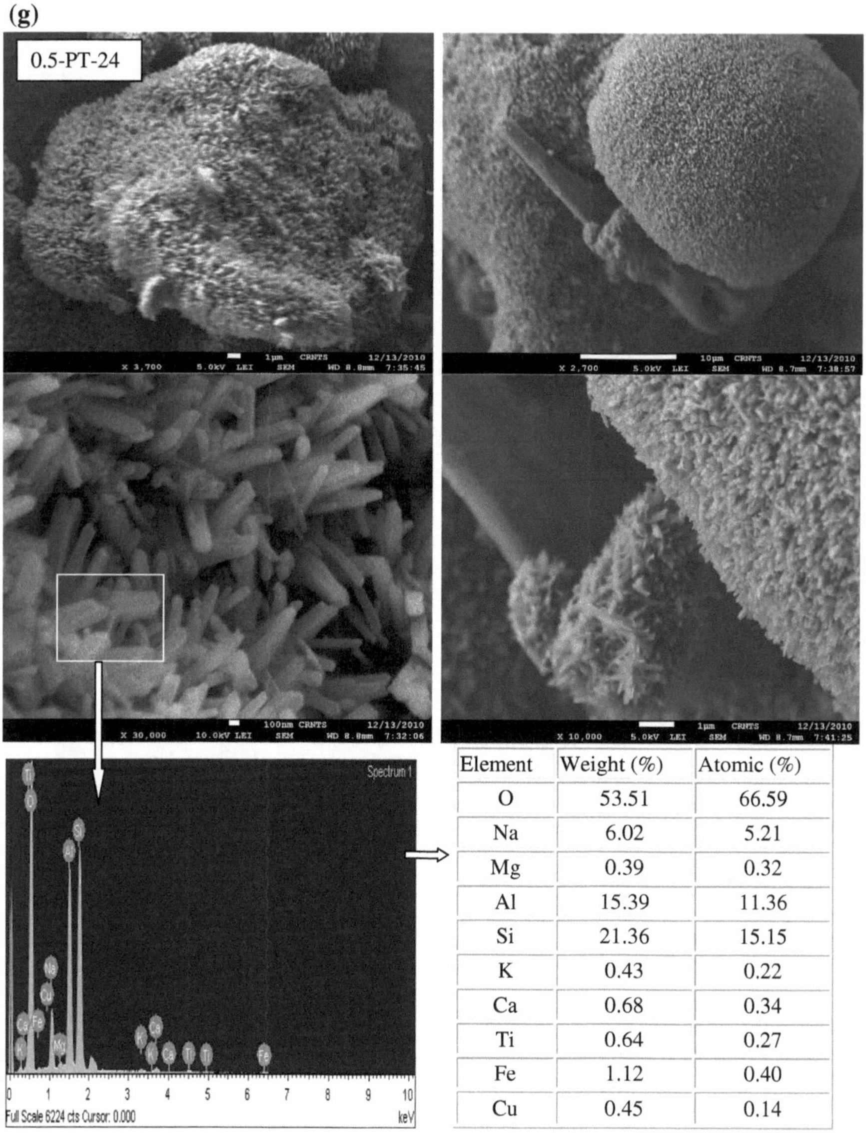

Element	Weight (%)	Atomic (%)
O	53.51	66.59
Na	6.02	5.21
Mg	0.39	0.32
Al	15.39	11.36
Si	21.36	15.15
K	0.43	0.22
Ca	0.68	0.34
Ti	0.64	0.27
Fe	1.12	0.40
Cu	0.45	0.14

Fig. 5.10 (continued)

Incidentally, these zeolites have been identified from the XRD diffractograms of the samples, as presented in Table 5.38. The mineralogical phase transition in the activated residues also reveals the higher zeolitization characteristics of the OHA for synthesizing a class of zeolites with a slightly higher *SAR* value (2.64 > *SAR* > 1.60), which is in close proximity to common fly ash zeolites, Faujasite and Analcime [47].

(h)

Element	Weight (%)	Atomic (%)
O	55.25	67.74
Na	10.22	8.72
Al	11.15	8.10
Si	20.29	14.17
K	0.46	0.23
Ca	0.79	0.39
Ti	0.40	0.17
Fe	0.80	0.28
Cu	0.62	0.19

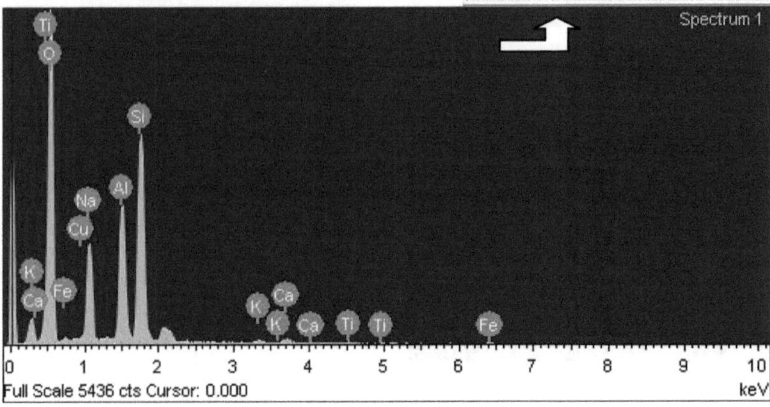

Fig. 5.10 (continued)

(i)

Element	Weight (%)	Atomic (%)
O	43.52	58.59
Na	3.87	3.62
Mg	0.24	0.21
Al	16.35	13.05
Si	26.06	19.99
K	0.56	0.31
Ca	3.10	1.67
Ti	1.95	0.88
Fe	4.34	1.67

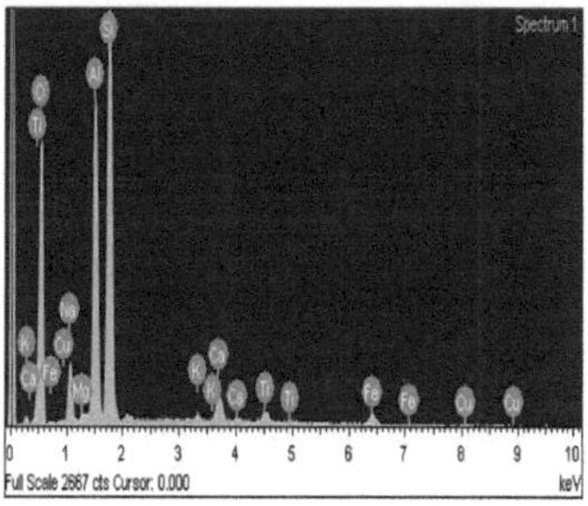

(j)

Element	Weight (%)	Atomic (%)
O	53.77	66.34
Na	11.45	9.83
Al	10.95	8.01
Si	20.93	14.71
K	0.42	0.21
Ca	0.29	0.14
Ti	0.49	0.20
Fe	0.70	0.25
Cu	0.99	0.31

Fig. 5.10 (continued)

(k)

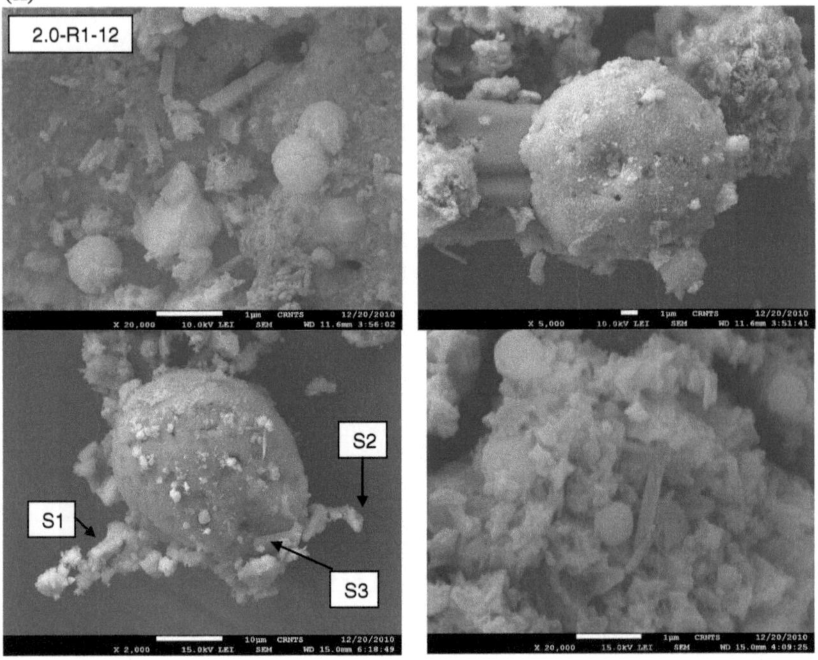

S1

Element	Weight (%)	Atomic (%)
O	61.31	73.46
Na	4.26	3.55
Al	6.28	4.46
Si	25.73	17.56
Ca	0.71	0.34
Ti	0.60	0.24
Fe	1.11	0.38

S2

Element	Weight (%)	Atomic (%)
O	63.17	73.65
Na	16.83	13.66
Al	9.67	6.68
Si	7.67	5.10
Ca	0.42	0.19
Ti	0.36	0.14
Fe	0.60	0.20
Cu	1.27	0.37

S3

Element	Weight (%)	Atomic (%)
O	56.06	67.52
Na	21.37	17.91
Al	12.31	8.79
Si	6.40	4.39
Ca	0.69	0.33
Ti	0.55	0.22
Fe	0.93	0.32
Cu	1.68	0.51

Fig. 5.10 (continued)

5.4.2.1 Variation of Cation Exchange Capacity (*CEC*)

The variation of *CEC* of the residues with respect to the molarity, M, is depicted in Fig. 5.19. It can be seen from this figure that the trends are not well defined, but the AHAs attain a higher *CEC* than the ALAs.

(l)

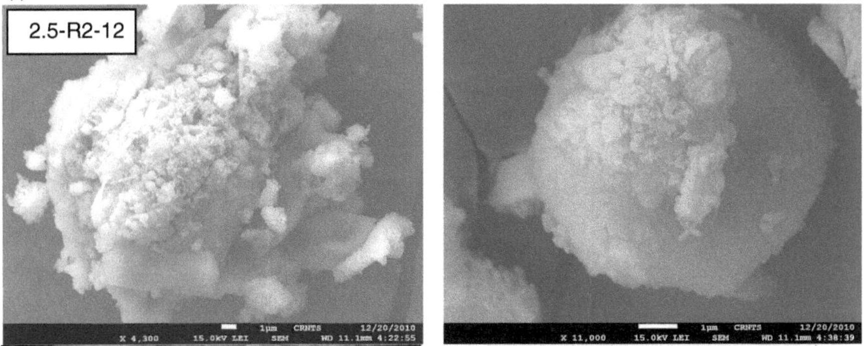

Element	Weight (%)	Atomic (%)
O	51.25	64.90
Na	7.52	6.63
Al	14.49	10.88
Si	21.73	15.67
Ca	0.87	0.44
Ti	0.61	0.26
Fe	1.48	0.54
Ni	1.03	0.36
Cu	1.02	0.32

Fig. 5.10 (continued)

This also confirms the higher zeolitization of the OHA, as compared to its counterpart. It should also be noted that some of the OHA residues (i.e., AHA3, AHA13 and AHA15) exhibit very high *CEC* values (275, 284 and 458 meq./100 g, respectively).

5.4.3 Correlation Between CEC *and* SAR

An attempt has been made to develop a correlation between the *CEC* and *SAR* of the residues, as depicted in Fig. 5.20. It is clear from this figure that the *CEC* decreases with an increase in *SAR*. However, the rate of this decrease (defined in terms of the slope of the relationship) is significantly steep for the AHA, as compared to the ALA. This also demonstrates that the OHA is a much more active and suitable material to synthesize a higher grade of cation exchanger (i.e., a zeolite)

(m)

Element	Weight (%)	Atomic (%)
O	48.69	62.19
Na	11.01	9.79
Al	12.03	9.11
Si	23.28	16.94
K	0.27	0.14
Ca	0.64	0.33
Ti	0.77	0.33
Fe	2.56	0.94
Cu	0.74	0.24

Fig. 5.10 (continued)

than the OLA. A combined trend for the ALA and AHA samples is also shown in the figure, which could be very useful for predicting the *CEC* of an activated fly ash if its *SAR* is known. Based on the *CEC* and *SAR* values of some common fly ash zeolites synthesized in this study, zeolites Na-P1 and Faujasite are marked above the general trend whereas Hydroxy-sodalite and Cancrinite are placed below the common trend (shown by dotted line), marked on Fig. 5.20.

(n)

Element	Weight (%)	Atomic (%)
O	55.02	65.53
Na	34.76	28.81
Al	2.63	1.86
Si	3.59	2.43
Ca	0.57	0.27
Fe	1.76	0.60
Cu	1.67	0.50

Element	Weight (%)	Atomic (%)
O	53.23	66.39
Na	9.62	8.35
Al	19.41	14.35
Si	12.50	8.88
Ca	0.82	0.41
Ti	0.72	0.30
Fe	3.70	1.32

(o-i)

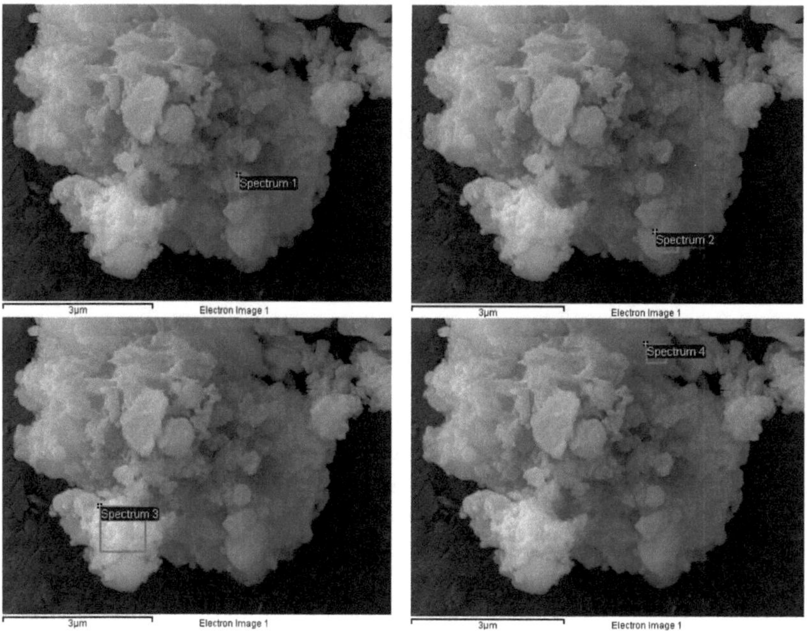

Fig. 5.10 (continued)

(o-ii)

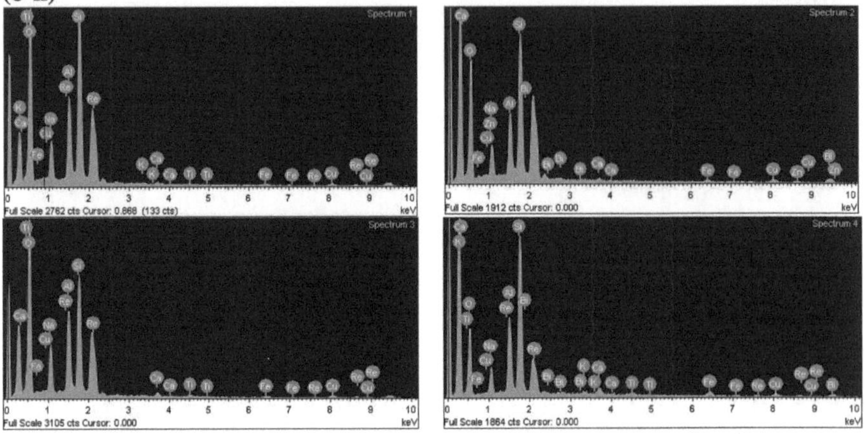

(o-iii)

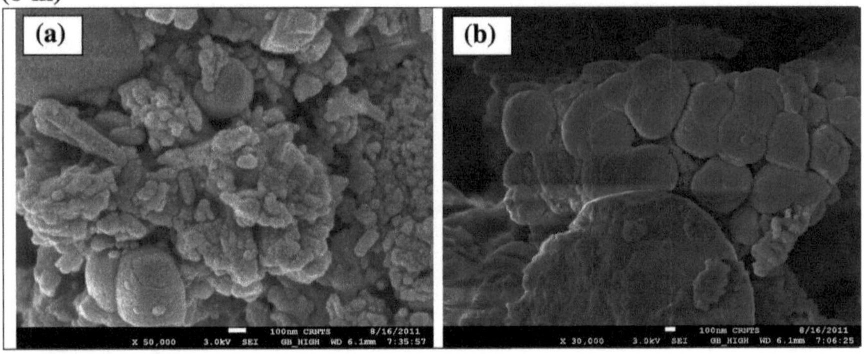

Fig. 5.10 (continued)

5.4.3.1 Mineralogical Alterations

Based on the X-ray diffractograms of both these ashes (OLA and OHA) and their activated residues, as depicted in Fig. 5.21a, b, results obtained after peak matching with the help of JCPDS data files [35] have been listed in Table 5.38. It can be deciphered from these results that the ALA samples contain fewer crystals of zeolite Na-P1 (refer Fig. 5.6). Most specifically, the residue ALA6 exhibits the presence of the zeolite Na-P1 for $M \leq 1$ and Hydroxy-sodalite for $M \geq 1$ [23]. On the other hand, Fig. 5.4b exhibits the peaks corresponding to the zeolites Na-P1 and Hydroxy-sodalite in the residue, AHA2. However, as listed in Tables 5.4 and 5.5, the zeolites, Cancrinite and Analcime, were also found in the residue AHA6. This

(p)

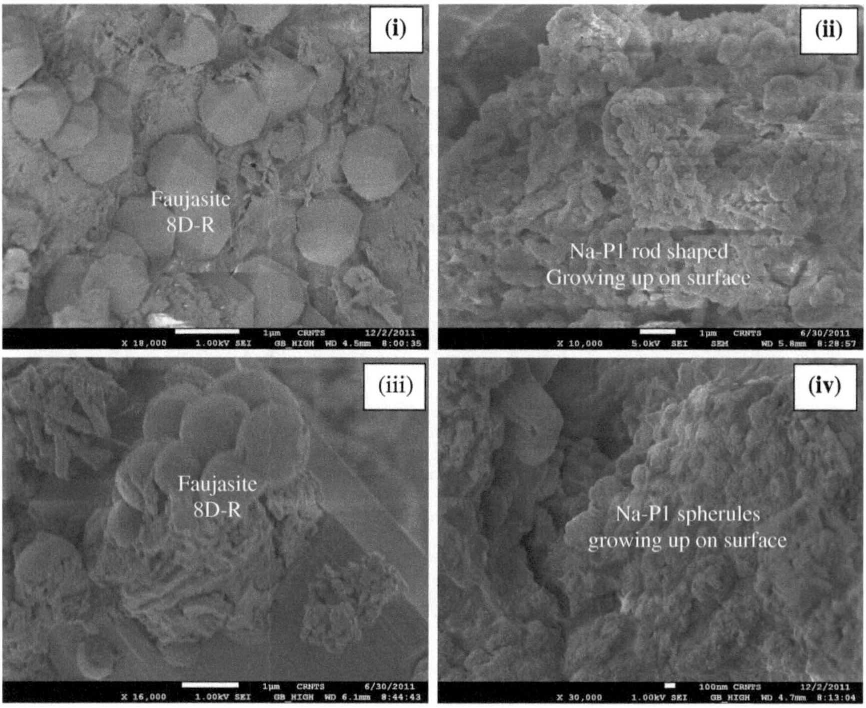

Fig. 5.10 (continued)

could be attributed to the disappearance of Mullite and hence the enhanced zeolitization of the hopper ash. Also, newly formed peaks ($2\theta°$ equal to 12.55° and 28.19°) exhibit an increase in the peak intensity with an increase in the molarity ($M \leq 1$) and activation time ($T \leq 24$ h).

From Table 5.38 it can be seen that the residue, AHA13, exhibits two new and small peaks of zeolite Analcime. Incidentally, these peaks and other peaks (present in the AHA15) disappear for a higher molarity and a longer activation time. As explained earlier, a high *CEC* (458 meq./100 g) of the residue AHA15, indicates a highly fibrous, porous and soft grade (low intensity peaks, refer Fig. 5.4b) of polycrystalline zeolites in the residues, as exhibited in their micrographs. Furthermore, the presence of some Quartz and Mullite in the residue AHA15 reveals the suitability of the hopper ash for the synthesis of superiorfly ash zeolites.

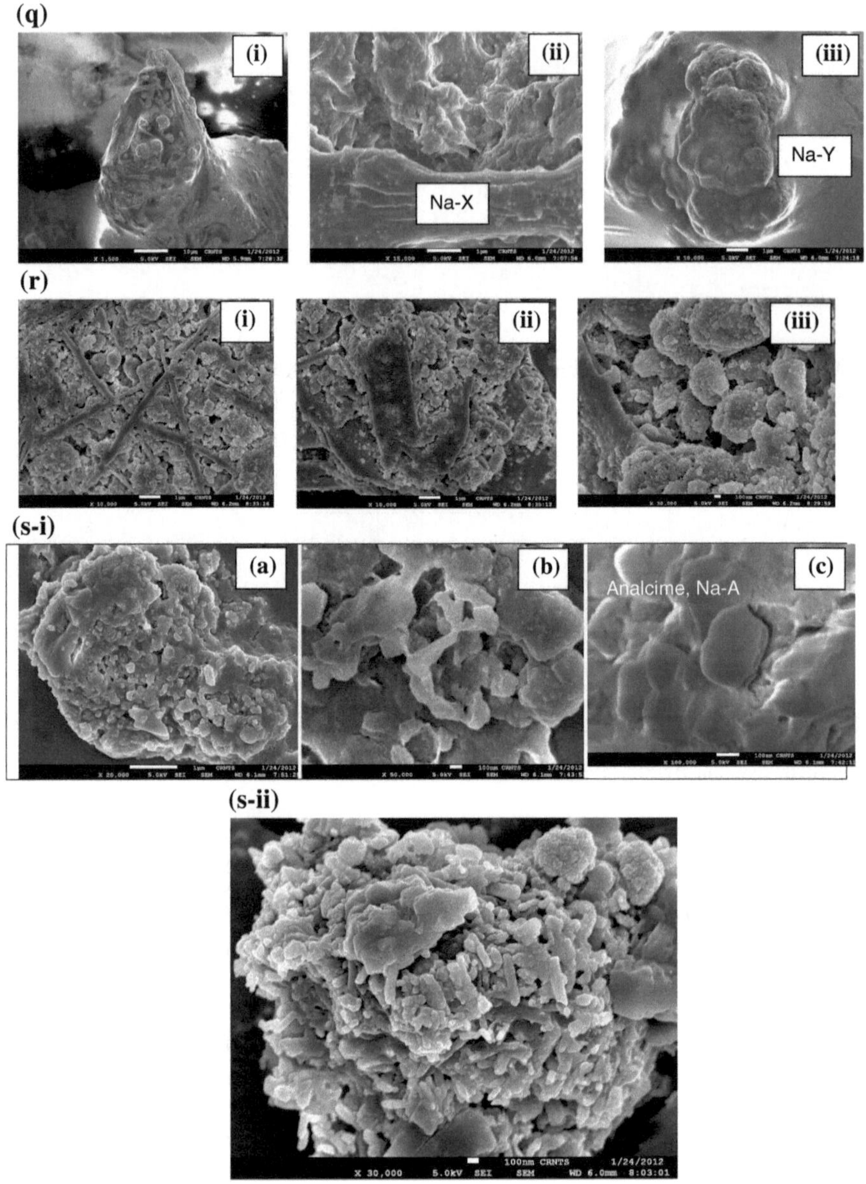

Fig. 5.10 (continued)

(t)

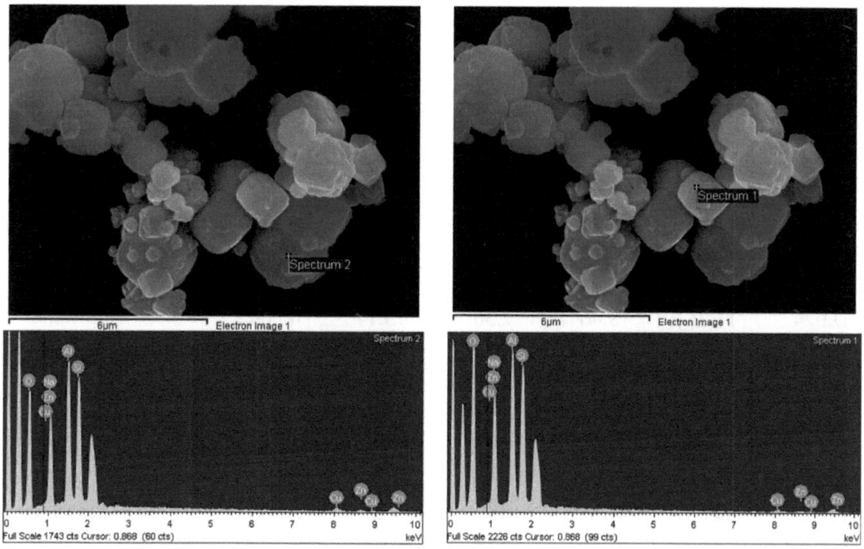

Element	Weight (%)	Atomic (%)
O	41.07	53.39
Na	14.45	13.07
Al	20.00	15.42
Si	24.47	18.12
Cu	0.00	0.00
Zn	0.00	0.00

Element	Weight (%)	Atomic (%)
O	45.35	57.54
Na	15.12	13.35
Al	18.00	13.54
Si	21.53	15.57
Cu	0.00	0.00
Zn	0.00	0.00

Fig. 5.10 (continued)

5.4.3.2 Morphological Alterations

From Fig. 5.22a, b it can be seen that both OHA and OLA samples are enriched with spheroidal and oval particles, which may correspond to glass and Quartz, respectively [10]. However, Fig. 5.22a exhibits some whitish spherules (zeolite

Table 5.35 Various elements of FAZs, 1.5-R2-24, dissolved in acidic solution

Sample	Concentration of elements in interacted acidic solution (ppm)						
	Fe	Ca	K	Na	Al	Mg	Si
ZD1-6	6.75	15.25	1.51	46.76	26.16	3.11	27.73
ZD1-12	6.49	9.41	1.67	42.23	22.89	1.55	26.28
ZD1-18	7.07	15.34	2.54	46.31	24.28	3.28	28.72
ZD1-24	7.38	15.46	1.45	45.18	25.03	3.35	29.64
ZD2-6	11.00	20.13	2.62	75.46	37.60	4.33	41.19
ZD2-12	11.62	20.20	3.00	80.61	39.53	4.59	43.46
ZD2-18	12.02	19.07	2.18	73.74	40.01	4.01	42.67
ZD2-24	11.76	18.40	2.36	75.14	45.00	3.93	42.93
ZD3-6	18.95	22.56	3.37	118.00	60.43	4.75	63.51
ZD3-12	18.30	21.85	3.38	119.00	54.85	4.91	61.56
ZD3-18	20.41	24.93	3.77	117.00	61.23	5.27	63.05
ZD3-24	20.39	24.31	3.72	116.00	61.52	5.32	62.85
ZD4-6	50.76	53.28	9.70	400.00	62.00	12.02	119.00
ZD4-12	54.45	55.01	9.81	256.00	61.50	12.63	118.00
ZD4-18	55.82	55.59	9.65	400.00	62.50	11.74	119.50
ZD4-24	60.72	53.99	10.11	412.00	63.00	12.20	120.00

Note ZD suffixed by 1, 2, 3 and 4 represent zeolite doses e.g. 0.6, 1.0, 1.66 and 5.0 g/l, respectively. A number after hyphen indicates the interaction time in hours

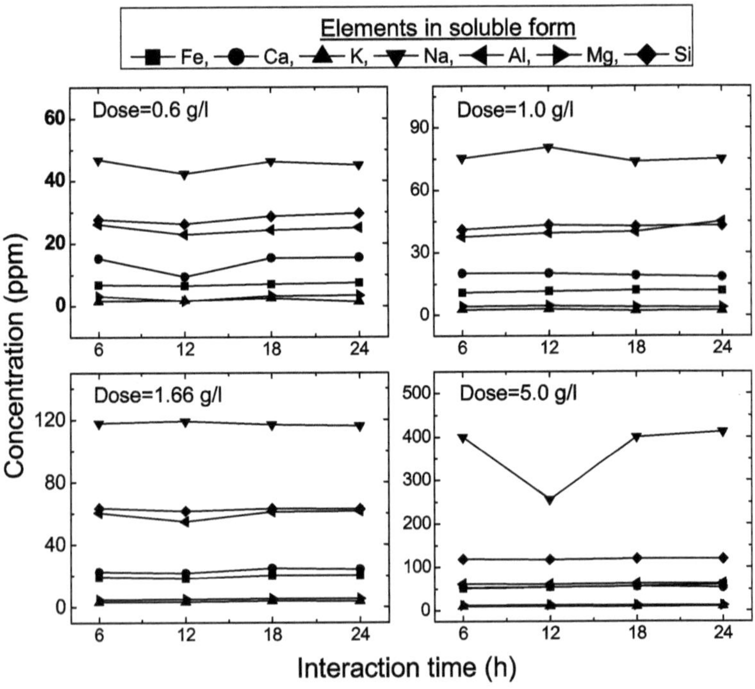

Fig. 5.11 Variation of dissolved elements of FAZ with zeolite-acid interaction time

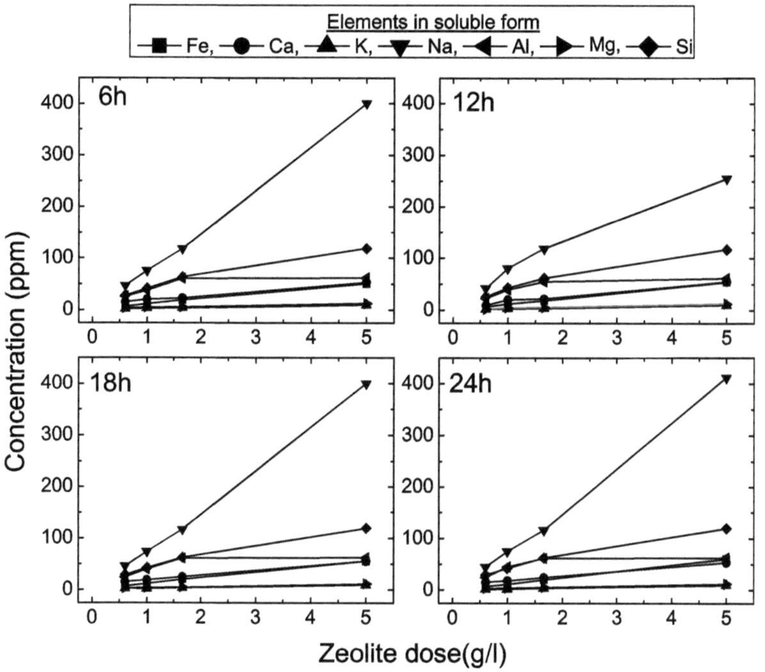

Fig. 5.12 Effect of FAZ dose on the concentration of dissolved elements of FAZs'

Fig. 5.13 Variation in response of the FAZ under thermal analyses. **a** TGA and **b** DTA

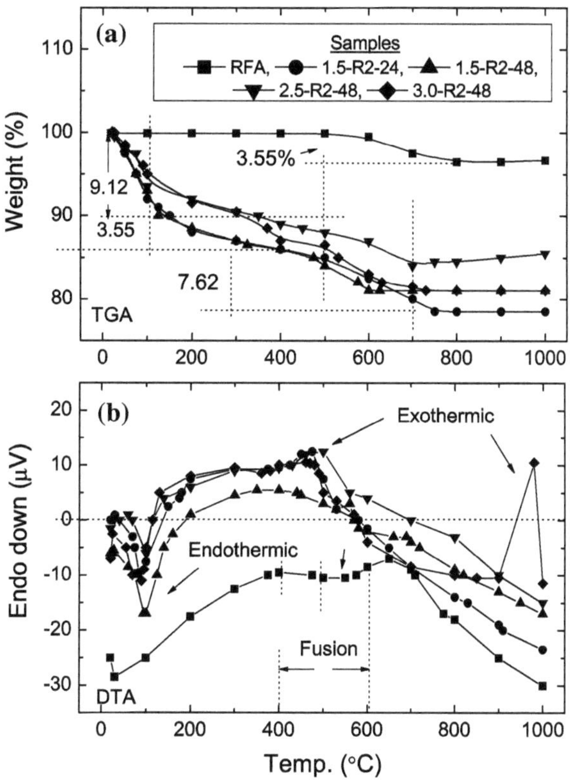

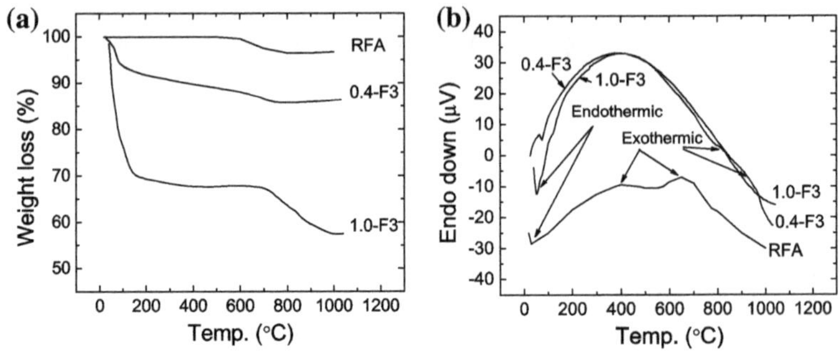

Fig. 5.14 Thermal response of the raw fly ash, fused fly ashes and the standard zeolite RZP by **a** TGA and **b** DTA analyses

Na-P1 with large surface pores) deposited on the OLA samples, due to the ash-water interaction taking place in the lagoon [46, 48, 49]. In contrast, the micrograph of the OHA (as depicted in Fig. 5.22b), reveals many smooth and spherical (Quartz, Mullite) to irregular particles (carbon, Hematite and/or Magnetite) as reported by Adamczyk and Bialecka [32] and Nugteren et al. [17].

It can also be noted from Fig. 5.23a that a spherical particle, in AHA6, has been fully covered with zeolites. The majority of the fibrous and ball-shaped products (see Fig. 5.23a, b) are zeolite Na-P1 and Hydroxy-sodalite, respectively [50, 51]. Also, orthorhombic and cubic morphologies (see Fig. 5.24c, d) correspond to another grade of Na-P1 (with improved *SAR* and *CEC*, as shown in Table 5.38 and Fig. 5.20) and Analcime. Also, some fine needle-shaped (Cancrinite) morphology [52] can also be seen in Fig. 5.24b, c. Above all, the cluster of whitish orthorhombic products formed in the AHA13 (see Fig. 5.24e, f) is exceptional and it confirms the significant zeolitization of the OHA samples. In addition, a large-scale development of the surface pores can be confirmed from Fig. 5.24a–f, which in turn supports the high *CEC* of the AHA samples. In agreement with the XRD patterns (see Fig. 5.21), the formation of the zeolitic morphology corresponds to the appearance of new mineral peaks. Moreover, the micrographs of the AHA15 (corresponding to 3.5 M NaOH and 36 h of treatment as exhibited in Fig. 5.24d), reveal the synthesis of the majority of zeolite Na-P1 with large inter-particle pores and a high *CEC* up to 458 meq./100 g (refer Figs. 5.20 and 5.24).

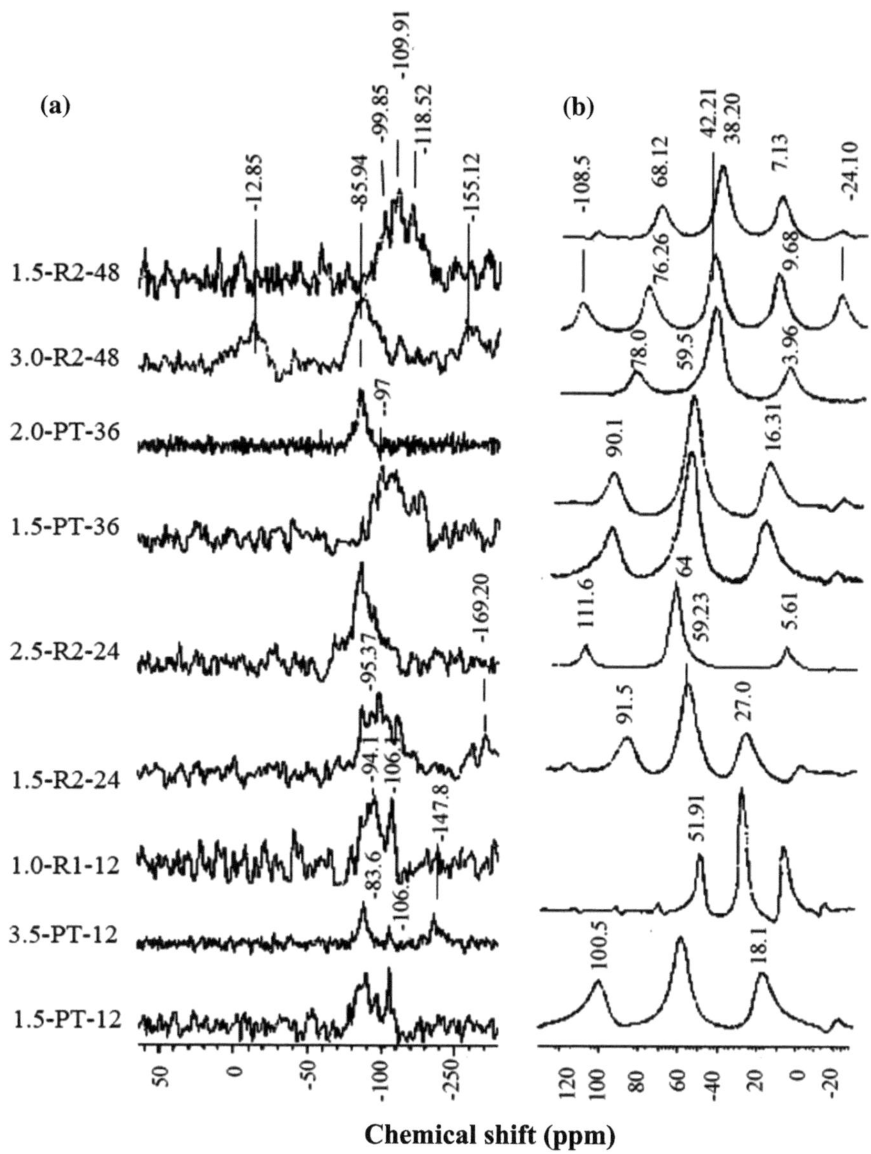

Fig. 5.15 NMR response (in *ppm*, as on *x-axis*) spectra of superior AAF for nuclei **a** ^{29}Si and **b** ^{27}Al

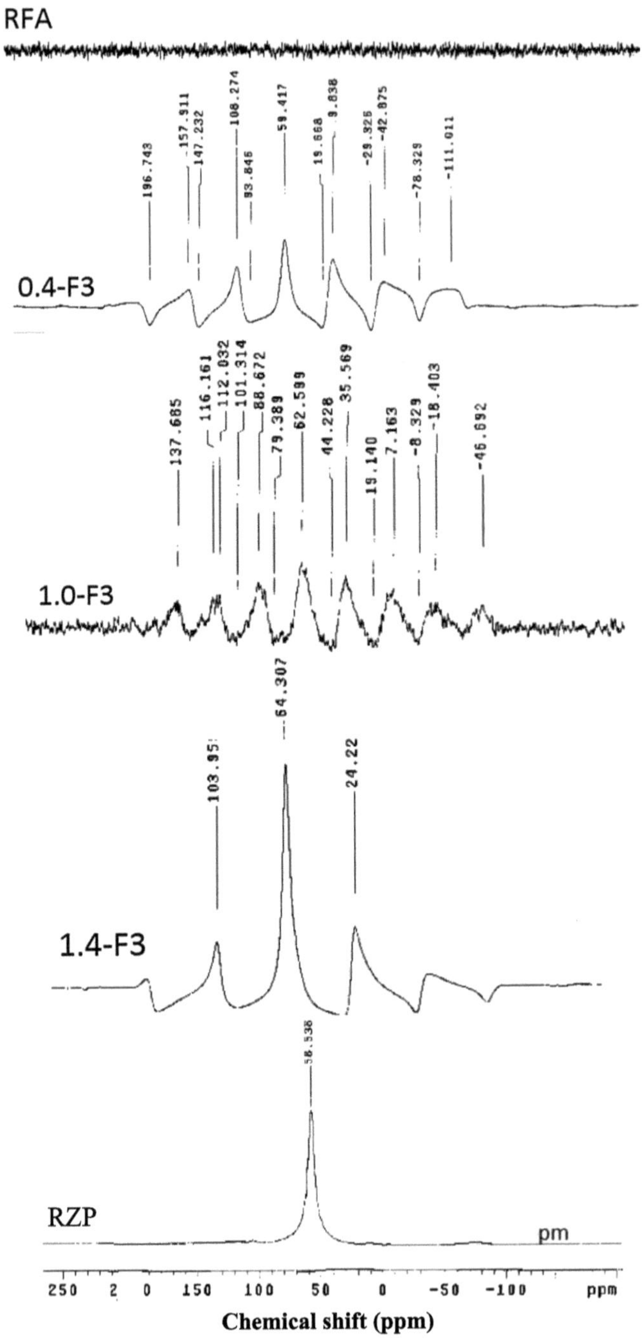

Fig. 5.16 ^{27}Al NMR spectra of the fly ash, fused residues and the reference zeolite 4A (RZP)

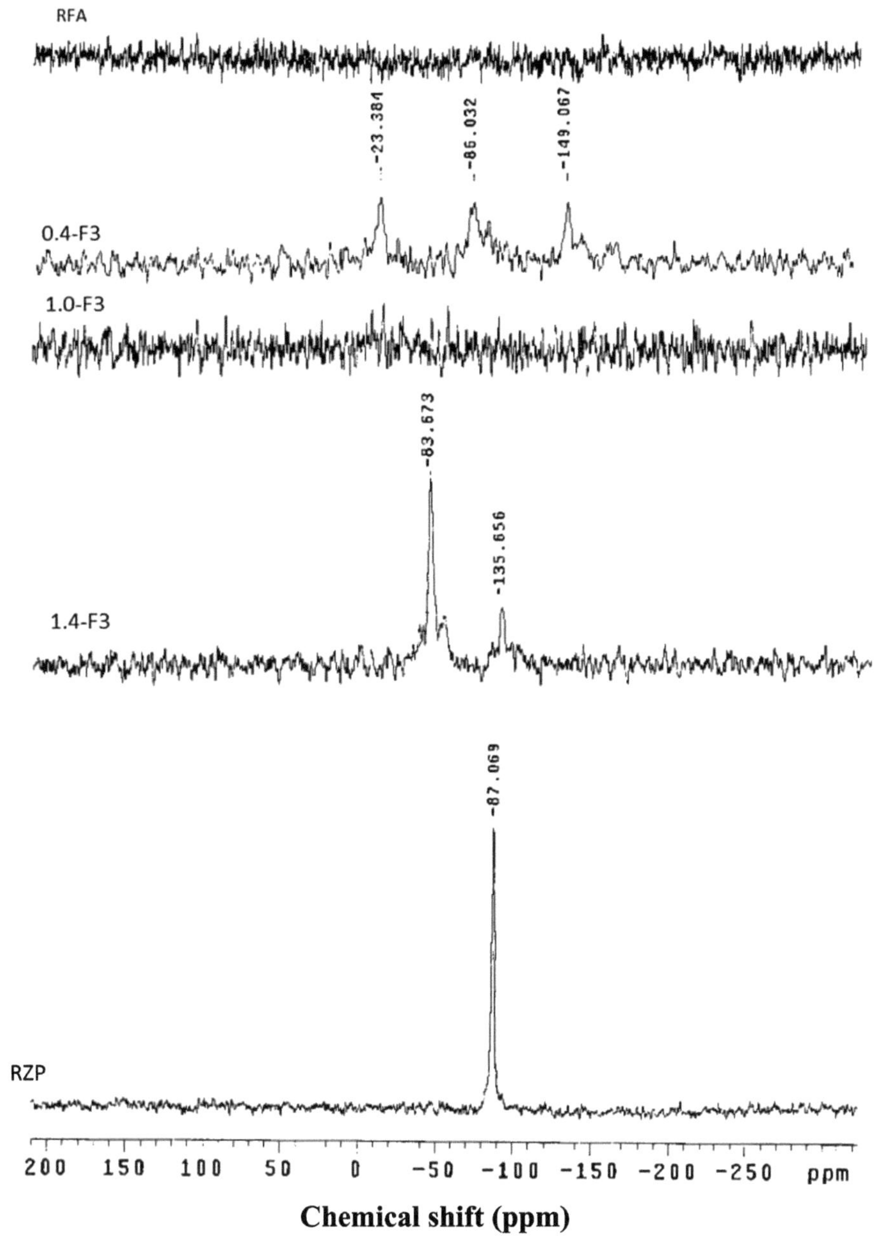

Fig. 5.17 ^{29}Si NMR spectra of the fly ash, fused residues and the reference zeolite 4A (RZP)

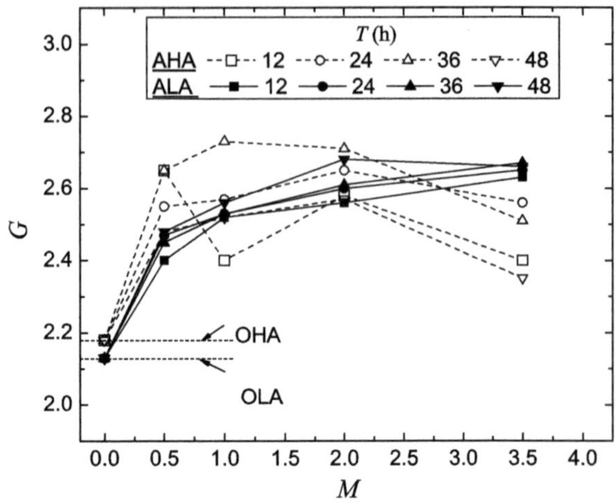

Fig. 5.18 Variation of *G* of AHA and ALA samples

Table 5.36 Chemical composition of the original fly ash

Oxide	% by weight	
	OHA	OLA
SiO_2	63.85	58.33
Al_2O_3	26.05	27.10
Fe_2O_3	5.14	5.02
Na_2O	0.06	0.22
K_2O	0.66	0.89
CaO	1.88	1.12
MgO	0.39	0.53
TiO_2	1.52	1.63
P_2O_5	0.16	0.24
$SiO_2 + Al_2O_3 + Fe_2O_3$	95.04	90.45

Table 5.37 Silica and alumina oxides and their ratio in the original and activated fly ashes (the residues)

Fly ash/Residue	Designation of the residue	SiO_2	Al_2O_3	*SAR*
		% by weight		
OHA	–	63.85	26.05	2.45
AHA	AHA1	57.73	27.90	2.06
	AHA2	48.74	21.70	2.24
	AHA3	52.05	24.00	2.16
	AHA4	48.74	18.40	2.64
	AHA5	52.56	20.88	2.51
	AHA6	55.15	24.64	2.23
	AHA7	36.96	18.76	1.97
	AHA8	51.41	23.56	2.18
	AHA9	36.77	19.10	1.92
	AHA10	38.67	26.00	1.48
	AHA11	36.77	19.10	1.92
	AHA12	41.7	17.66	2.36
	AHA13	37.93	23.65	1.60
	AHA14	35.48	19.76	1.79
	AHA15	43.83	17.74	2.47
	AHA16	28.12	14.30	1.96
OLA	–	58.33	27.10	2.15
ALA	ALA1	64.2	26.77	2.39
	ALA2	62.29	29.27	2.12
	ALA3	59.78	30.87	1.93
	ALA4	58.55	30.04	1.94
	ALA5	61.53	30.84	1.99
	ALA6	58.94	31.12	1.89
	ALA7	54.89	30.47	1.80
	ALA8	57.03	32.76	1.74
	ALA9	58.01	30.90	1.87
	ALA10	56.98	32.16	1.77
	ALA11	56.06	32.53	1.72
	ALA12	53.94	34.14	1.57
	ALA13	48.76	30.99	1.57
	ALA14	43.27	32.29	1.34
	ALA15	42.12	32.48	1.29
	ALA16	40.59	32.20	1.26

– denotes not any

Table 5.38 Minerals identified in the fly ash and superior residues

Mineral	OHA	OLA	AHA2	AHA6	ALA6	ALA9	ALA10	ALA11	AHA13	AHA15	ALA16
Quartz	✓✓✓		✓✓						✓	–	✓✓
Mullite	✓✓		✓	–	✓				–		✓
Na-P1 zeolite	–	✓	✓	✓✓		✓		–	✓	✓✓	–
Hydroxy-sodalite	–			–	✓	–			–		✓
Cancrinite			–	✓	–				–		
Analcime				–					✓		–

Note predominant (✓✓✓), present (✓✓), traces (✓) and absent (–)

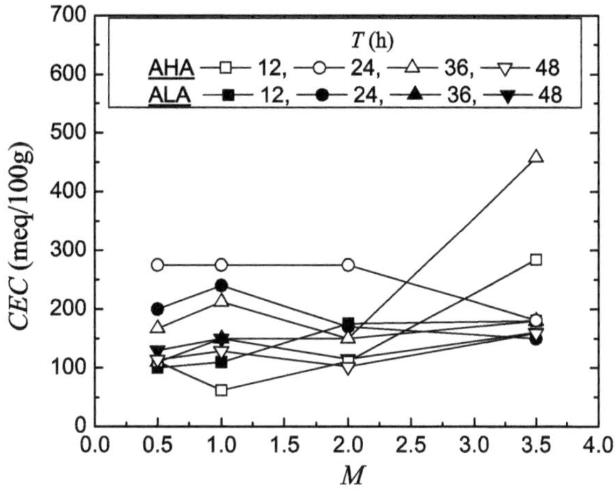

Fig. 5.19 Variation of *CEC* of the AHA and the ALA samples

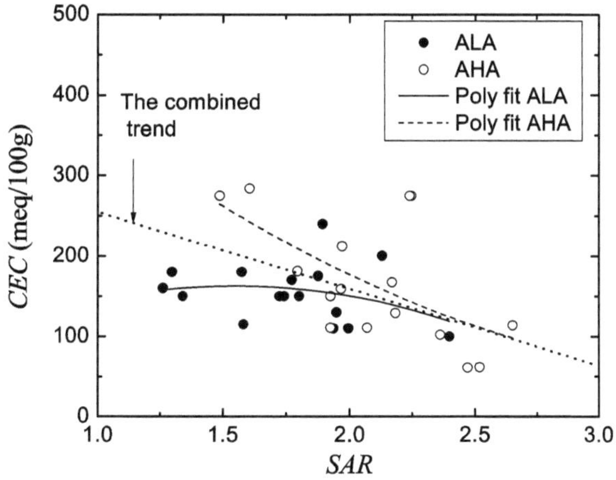

Fig. 5.20 Correlation between *CEC* and *SAR* of the residues

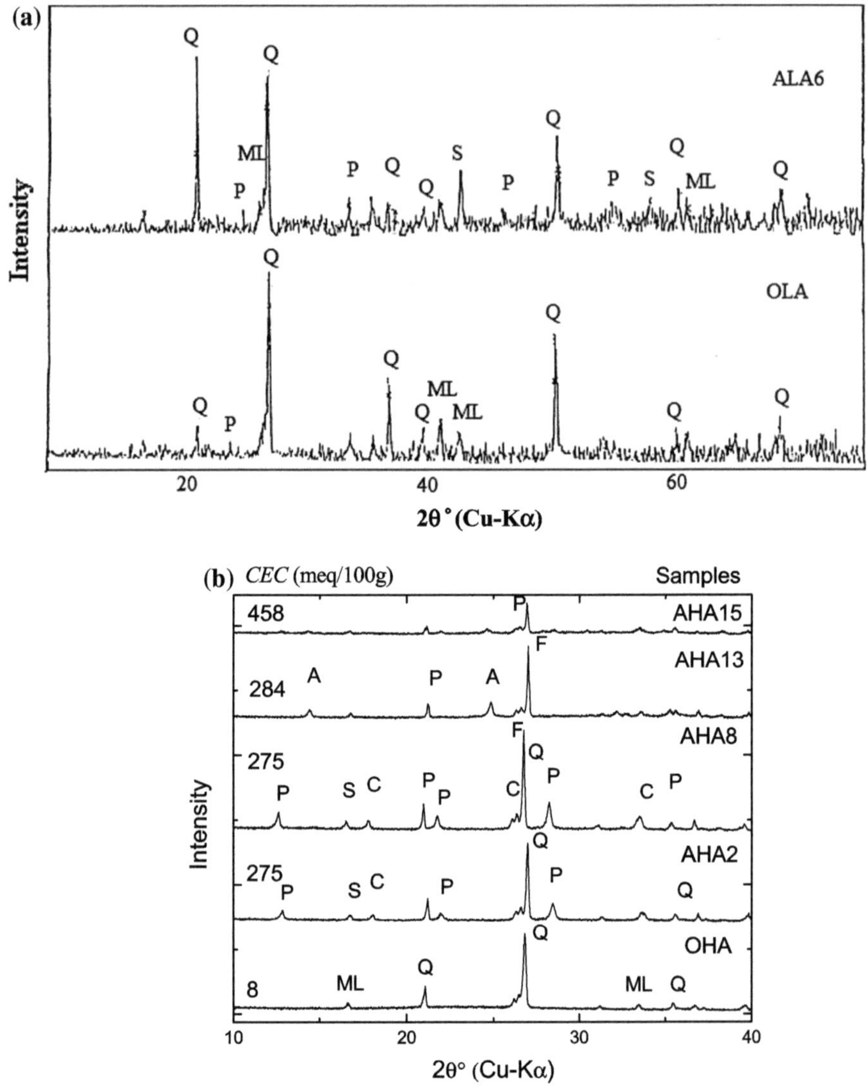

Fig. 5.21 **a** XRD patterns of the OLA and the superior ALA samples, where Q, ML, P and S represent Quartz, Mullite, zeolite Na-P1 and Hydroxy-sodalite, respectively, and θ is the angle of scattering of X-ray. **b** XRD patterns of the OHA and the superior AHA samples (where Q, ML, P, C, S and A represent Quartz, Mullite, zeolite Na-P1, Cancrinite, Hydroxy-sodalite and Analcime, respectively)

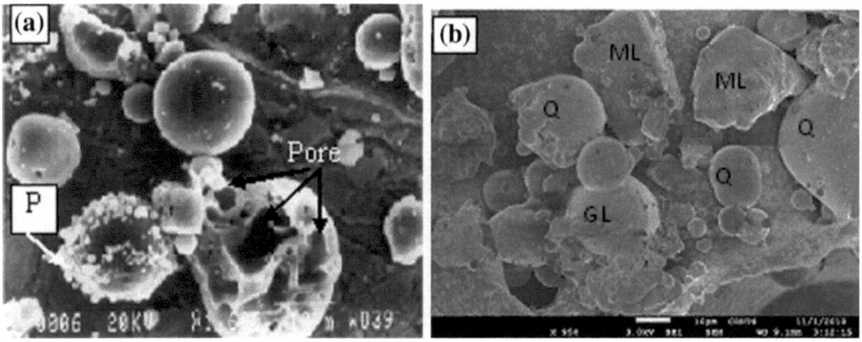

Fig. 5.22 Micrographs of the samples. **a** OLA and **b** OHA, where Q, ML, GL and P designate Quartz, Mullite, glass and zeolite Na-P1, respectively

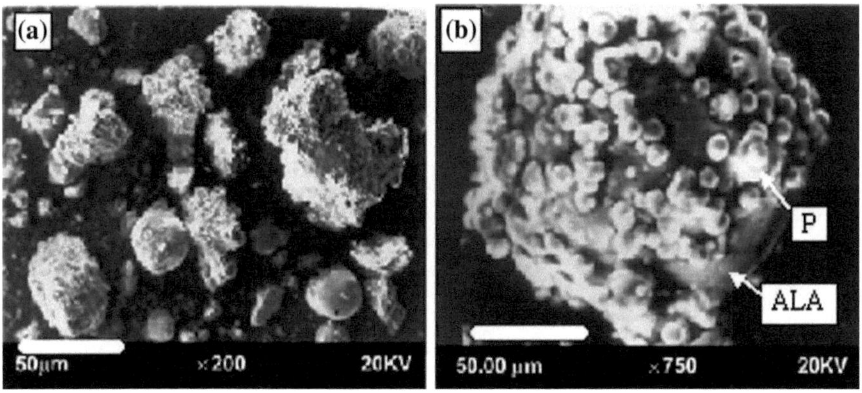

Fig. 5.23 Micrographs of the superior residue, ALA6, at different magnifications **a** ×200 and **b** ×750

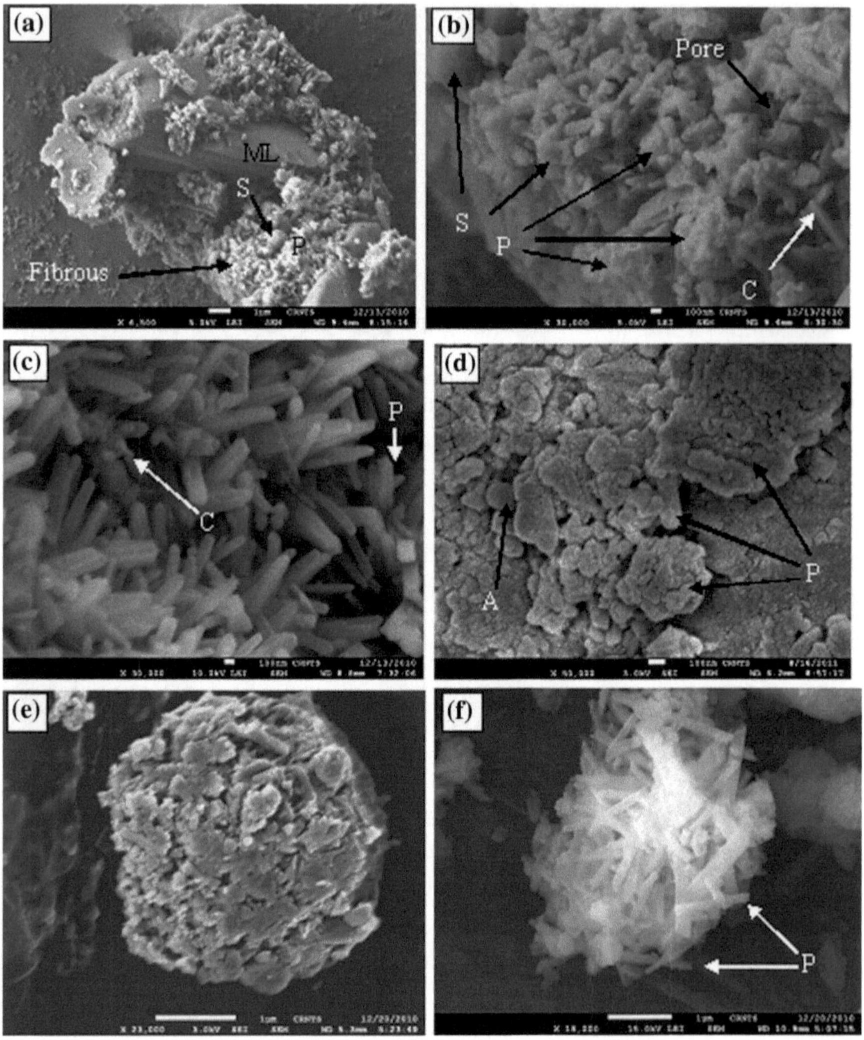

Fig. 5.24 **a** AHA6, ×6500, **b** AHA6, ×30,000, **c** AHA2, ×30,000, **d** AHA15, ×50,000, **e** AHA13, ×23,000 and **f** AHA13, ×18,000, where ML, P, C, S and A designate the Mullite, zeolites Na-P1, Cancrinite, Hydroxy-sodalite and Analcime, respectively

5.5 Conclusions

Based on the findings presented in this chapter, it can be opined that the hydrothermal treatment of the fly ash collected from the hopper yields much better zeolites as compared to its counterpart (the fly ash collected from the lagoon). The three-step activation of the hopper ash with NaOH results in several changes (viz., minor variation in pH and reduction in electrical conductivity) in the supernatant. The three-step activation by hydrothermal technique causes reduction in Si and Al contents of the supernatant, obtained after the treatments. This process yields improved grade of the fly ash zeolites with higher *CEC* and specific surface area, enhanced specific gravity, nano sized fine particles of zeolites, micro-sized new pores. It has also been observed that the-three step activation by fusion techniques also effective for zeolitization of the fly ash, however, presence of inactivated impurities in the fusion residues makes them inferior residues as compared to the hydrothermal residues.

References

1. Lin, C.F., His, H.C.: Resource recovery of waste fly ash: synthesis of zeolite-like materials. Environ. Sci. Technol. **29**, 1109–1117 (1995)
2. Scott, J., Guang, D., Naeramitmarnsuk, K., Thabuot, M., Amal, R.: Zeolite synthesis from coal fly ash for the removal of lead ions from aqueous solution. J. Chem. Technol. Biotechnol. **77**, 63–69 (2001)
3. Juan, R., Hernández, S., Andrés, J.M., Ruiz, C.: Synthesis of granular zeolitic materials with high cation exchange capacity from agglomerated coal fly ash. Fuel **86**, 1811–1821 (2007)
4. Wałek, T.T., Saito, F., Zhang, Q.: The effect of low solid/liquid ratio on hydrothermal synthesis of zeolites from fly ash. Fuel **87**, 194–199 (2008)
5. Kim, W., Seung, H., Ahn, B.J.: Synthesis of Na-P1 zeolite from coal fly ash. J. Indus. Eng. Chem. **3**(3), 185–190 (1997)
6. Hollman, G.G., Steenbruggen, G., Janssen, J.M.: A two-step process for the synthesis of zeolites from coal fly ash. Fuel **78**, 1225–1230 (1999)
7. Somerset, V.S., Petrik, L.F., White, R.A., Klink, M.J., Iwuoha, E.: The use of X-ray fluorescence (XRF) analysis in predicting the alkaline hydrothermal conversion of fly ash precipitates into zeolites. Talanta **64**, 9–14 (2004)
8. Jha, B., Singh, D.N.: A review on synthesis, characterization and industrial application of flyash zeolites. J. Mater. Edu. **33**(1–2), 65–132 (2011)
9. Jha, B., Singh, D.N.: Zeolitizationcharacteristics of a fly ash from wet- and dry- disposal systems. Acta. Geotech. Slov. **2**, 63–71 (2012)
10. Jha, B., Padmakumar, G.P., Singh, D.N., Iyer, K.: Synthesis of zeolites by fly ash alkali interaction. IN: Proceedings of Indian Geotechnical Conference, Cochi, India, pp. 1089–1092 (2011)
11. Jha, B., Singh, D.N.: Synthesis of higher grade fly ash zeolites X from fly ash via three-step fusion. Mater. Perform. Charact. **2**(1), 1–12 (2013)
12. Jha, B., Singh, D.N.: Formation of meso- and micro-pores in fly-ash zeolites using three step activation. Acta. Geotech. Slov. **1**, 63–69 (2014)
13. Jha, B., Singh, D.N.: A three step process for purification of fly ash zeolites by hydrothermal treatment. App. Clay. Sci. **90**, 122–129 (2014)
14. Jha, B., Singh, D.N.: Quantification of transitions occurring in a hydrothermally activated fly ash. Mater. Perform. Charact. **3**(1), 239–254 (2014)
15. Jha, B., Koshy, N., Singh, D.N.: Establishing two-stage interactions between fly ash and NaOH by X-ray and infrared analyses. Front. Environ. Sci. Eng. **9**(2), 216–221 (2015)
16. Inada, M., Eguchi, Y., Enomoto, N., Hojo, J.: Synthesis of zeolite from coal fly ashes with different silica-alumina composition. Fuel **84**, 299–304 (2005)
17. Nugteren, H.W., Moreno, N., Sebastia, E., Querol, X.: Determination of the available Si and Al from coal fly ashes under alkaline conditions with the aim of synthesizing zeolites products.

International Ash Utilization Symposium, Centre for Applied Energy Research, University of Kentuchy, Paper No. 71 (2001)
18. Ojha, K., Pradhan, N.C., Samanta, A.N.: Zeolite from fly ash synthesis and characterization. Bull. Mater. Sci. **27**(6), 555–564 (2004)
19. Shigemoto, N., Hayashi, H.: Selective formation of Na-X zeolite from coal fly ash by fusion with sodium hydroxide prior to hydrothermal reaction. J. Mater. Sci. **28**, 4781–4786 (1993)
20. Berkgaut, V., Singer, A.: High capacity cation exchanger by hydrothermal zeolitization of coal fly ash. App. Clay. Sci. **10**, 369–378 (1996)
21. Rungsuk, D., Apiratikul, R., Pavarajarn, V., Pavasant, P.: Zeolite synthesis from coal-fired power plant by fusion method. In: The 2nd International Conference on Sustainable Energy and Environment, 21–23 November, Bangkok, Thailand (2006)
22. Ma, W., Brown, P.W., Komarneni, S.: Characterization and cation exchange properties of zeolites synthesized from fly ashes. J. Mater. Res. **13**(1), 3–7 (1998)
23. Kolay, P.K., Singh, D.N.: Physical, chemical, mineralogical and thermal properties of cenospheres from a ash lagoon. Cem. Conc. Res. **31**(4), 539–542 (2001)
24. Kolay, P.K., Singh, D.N.: Effect of zeolitization on physic-chemico-mineralogical and geotechnical properties of the lagoon ash. Canad. Geotech. J. **38**(5), 1105–1112 (2001)
25. Fernandez, P.C., Galiano, Y.L., Rodrıguez, P.M.A., Vale, J., Querol, X.: Utilisation of zeolitised coal fly ash as immobilising agent of a metallurgical waste. J. Chem. Technol. Biotechnol. **77**, 305–310 (2002)
26. ASTM D 5550-00 Standard Test: method for specific gravity of soil solids by helium gas pycnometer, Annual Book of ASTM Standard, 04.08, ASTM, Philadelphia, USA
27. Cerato, A.B., Lutenegger, A.J.: Determination of surface area of fine-grained soils by Ethylene Glycol Monoethyle Ether (EGME) method. Geotech. Test. J. **25**(3), 1–7 (2002)
28. Derkowski, A., Franus, W., Waniak-Nowicka, H., Czimerova, A.: Textural properties versus *CEC* and EGME retention of Na-X zeolite prepared from fly ash at room temperature. Inter. J. Miner. Process. 82, 57–68 (2006)
29. ASAP 2020: Accelerated surface area and porosimetry system. Operator's Manual V3.0, Micromeritics Instrument Corporation, 2004–2006
30. Gregg, S.J., Sing, K.S.W.: Adsorption, Surface area and Porosity, 2nd ed. Academic Press Inc, London (1982)
31. ISIRC World soil information data base. Wageningen UR (1992)
32. Adamczyk, Z., Bialecka, B.: Hydrothermal synthesis of zeolites from polish coal fly ash. Pol. J. Environ. Stud. **14**(6), 713–719 (2005)
33. Somerset, V.S., Petrik, L.F., White, R., Klinka, M.J., Key, D., Iwuoha, E.I.: Alkaline hydrothermal zeolites synthesized from high SiO_2 and Al_2O_3 co-disposal fly ash filtrates. Fuel **84**, 2324–2329 (2005)
34. Querol, X., Moreno, N., Alastuey, A., Juan, R., Andres, J.M., Lopez-Soler, A., Ayora, C., Medinaceli, A., Valero, A.: Synthesis of high ion exchange zeolites from coal fly ash. Geologica. Acta. **5**(1), 49–57 (2007)
35. JCPDS: Joint committee on powder diffraction standards, Philadelphia-19103 (1994)
36. Rayalu, S., Meshram, S.U., Hasan, M.Z.: Highly crystalline faujasitic zeolites from fly ash. J. Hazard. Mater. **B77**, 123–131 (2000)
37. Rayalu, S.S., Udhoji, J.S., Meshram, S.U., Naidu, R.R., Devotta, S.: Estimation of crystallinity in flyash-based zeolite-A using XRD and IR spectroscopy. Curr. Sci. **89**(12), 2147–2151 (2005)
38. Elzea, J.M., Rice, S.B.: TEM and X-ray diffraction evidence for cristobalite and tridymite stacking sequences in opal. Clays Clay Miner. **44**, 492–500 (1996)
39. Cullity, B.D., Stock, S.R.: Elements of X-ray Diffraction, 3rd Edn. Prentice Hall PTR (2001)
40. Bakoyannakis, D.N., Deliyanni, E.A., Zouboulis, A.I., Matis, K.A., Nalbandian, L., Kehagias, T.: Akaganeite and goethite-type nanocrystals: synthesis and characterization. Micropor. Mesopor. Mater. **59**, 35–42 (2003)

41. Baranova, E.A., Page, Y.L., Ilin, D., Bock, C., MacDougall, B., Mercier, P.H.J.: Size and composition for 1–5 nmϕPtRu alloy nano particles from Cu Kα X-ray patterns. J. Alloys. Compd. **471**, 387–394 (2009)
42. Mimura, H., Yokota, K., Akiba, K., Onodera, Y.: Alkali hydrothermal synthesis of zeolites from coal fly ash and their uptake properties of Cesium ion. J. Nuc. Sci. Technol. **38**(9), 766–772 (2001)
43. Criado, M., Fernandez-Jimenez, A., Palomo, A.: Alkali activation of fly ash: Effect of the SiO_2-Na_2O ratio, Part I: FT-IR study. Micropor. Mesopor. Mater. **10**, 180–191 (2007)
44. El-Naggar, M.R., El-Kamash, A.M., El-Dessouky, M.I., Ghonaim, A.K.: Two-step method for preparation of NaA-X zeolite blend from fly ash for removal of cesium ions. J. Hazard. Mater. **154**, 963–972 (2008)
45. Iyer, R.S., Scott, J.A.: Power station fly ash-a review of value-added utilization outside of the construction industry. Res. Conserv. Rec. **31**, 217–228 (2001)
46. Kolay, P.K., Singh, D.N.: Characterization of alkali activated lagoon ash and its application for heavy metal retention. Fuel **8**, 483–489 (2002)
47. Lobo, R.F.: Introduction to Structural Chemistry of Zeolites. Marcel Dekker Inc, Delaware, USA (2003)
48. Kolay, P.K., Singh, D.N., Murti, M.V.R.: Synthesis of zeolites from lagoon ash. Fuel **80**, 739–745 (2001)
49. Fazio, A.D., Brotzu, P., Ghiara, M.R., Fercia, M.L., Lonis, R., Sau, A.: Hydrothermal treatment at low temperature of Sardinian clinoptilolite bearing ignimbrites for increasing cation exchange capacity. Periodico. Di. Mineral. **77**, 79–91 (2008)
50. Rios, C.A.R., Williams, C.D., Roberts, C.L.: A comparative study of two methods for the synthesis of fly ash-based sodium and potassium type zeolites. Fuel **88**, 1403–1416 (2009)
51. Zeng, R., Umana, J.C., Querol, X., Lopez, S.A., Plana, F., Zhuang, X.: Zeolite synthesis from a high Si-Al fly ash from east China. J. Chem. Technol. Biotechnol. **77**, 267–273 (2002)
52. Wang, H., Pinnavaia, J.: MFI zeolite with small and uniform intracrystalmesopores. Angew. Chem. Int. Ed. **45**, 7603–7606 (2006)

Chapter 6
Major Findings of the Three-Step Activation Technique

Abstract The present chapter deals with the inferences derived from the "three-step activation" of the hopper ash, which has been ascertained to be the superior ash over lagoon ash, as described in Chap. 5. This technique has been found to be superior over conventional hydrothermal technique for obtaining the residues with high cation exchange values (viz., the higher grade zeolites, Na-P1, Hydroxy-sodalite, Faujasite, Cancrinite, Na-A). This technique has been established to be highly effective for purification of fly ash zeolites as well. Attempts have been made in the following to discuss about the quantification of transitions occurring into the filtrate and the macro to micro pores of the residues obtained from the three-step activation. Furthermore, efforts have been made to monitor the three-step activation of the fly ash by fusion method to synthesize high grade zeolite X. The final inference of this study is that the fly ash zeolites synthesized by following the hydrothermal technique is superior, as compared to the products of fusion technique.

Keywords Fly ash · Three step activation · Hydrothermal · Fusion · Zeolite · Quantification · Transition · Purification · Pores

6.1 Synthesis of Higher Grade Zeolites from Hopper Fly Ash

Synthesis of zeolites from fly ash by resorting to hydrothermal treatments of the (i) fly ash-NaOH mixture, in one step and the (ii) supernatant, the liquid by-product of such treatment, and $NaAlO_2$ mixture, in two steps, is well established. However, the presence of residual fly ash in such zeolites is responsible for their impurities (i.e., presence of Quartz and Mullite) and their low grade, in terms of the cation-exchange capacity and crystallinity. To overcome these issues, a novel methodology, "*three-step activation (TSA) of the fly ash by hydrothermal method*" has been developed and its details are presented in this section. It has been demonstrated that this method results in high grade polycrystalline zeolites, which

B. Jha and D.N. Singh, *Fly Ash Zeolites*, Advanced Structured Materials 78,
DOI 10.1007/978-981-10-1404-8_6

are significantly porous and grown up crystals, aluminated and high grade of cation-exchangers also conforming to a commercial zeolite Na-P1 (JCPDS code 39-0219) [1].

6.1.1 *Characteristics of the Filtrates*

The TSA results in an enhanced zeolitization of the fly ash, as affirmed from the changes in filtrates ($pH > 10$, refer Fig. 6.1a–d), indicates that Steps-2 and 3 of the treatment have minor effect on pH (also refer Tables 5.4 and 5.5). It is apparent from Fig. 6.1e–h and Tables 5.6 and 5.7 that electrical conductivity, EC, of the filtrates significantly decreases with increase in T_c from Step-1 to Step-3 of the TSA. Thus a continuous loss in EC can be attributed to incorporation of Na^+ (in majority) and other ions (viz., Ca^{2+}, K^+, Mg^{2+}) with $[SiO_4]^{4-}$ and $[AlO_4]^{5-}$ tetrahedra present in the residues.

In addition, Fig. 6.2a–h (refer Tables 5.8, 5.9, 5.10 and 5.11) reveals that these filtrates exhibit high concentration of Si^{4+} and low concentration of Al^{3+}, randomly, with increase in cumulative activation time, T_c, and molarity (M) of NaOH, at the end of Step-1 treatment (designated as PT). These findings are in agreement with those reported by earlier researchers [2] and can be attributed to (i) significant dissolution of soluble ingredients (viz., Quartz and amorphous glass) of the fly ash in the NaOH solution and (ii) relatively less dissolution of Al, bearing crystalline minerals like Mullite. It can also be observed from these figures that there is a significant reduction in concentration of Si^{4+} and Al^{3+} in the filtrates, which were obtained after recycling steps, R1 and R2. This could be attributed to (i) the interaction between the residual fly ash (i.e., the fly ash which did not get activated) and the filtrate and (ii) unification of the soluble ions in the filtrate which leads to the formation of complex ions, which may precipitate and hence may get mixed up with the residues during prolonged storage.

6.1.2 *Characteristics of the Hopper Fly Ash and Residues*

6.1.2.1 Bulk Chemical Composition of AAF Obtained from TSA

It can be observed from Tables 5.20, 5.21, 5.22, 5.23 and 5.36 that the sum total of major oxides ($Al_2O_3 + SiO_2 + Fe_2O_3$) present in the raw fly ash, RFA (i.e., the original hopper ash) is about 95 % and hence it can be characterized as Class-F fly ash (ASTM C618-08) [3]. It can be noted from Fig. 6.3 and Tables 5.20, 5.21, 5.22 and 5.23 that activation of this fly ash with NaOH results in loss of SiO_2 (in majority) and Al_2O_3 (in traces) in the residues (the alkali activated fly ash, AAF), which is represented as the second data point, from the origin, on the duration axis, for all the durations. Incidentally, reduction in the value of oxides increases with an

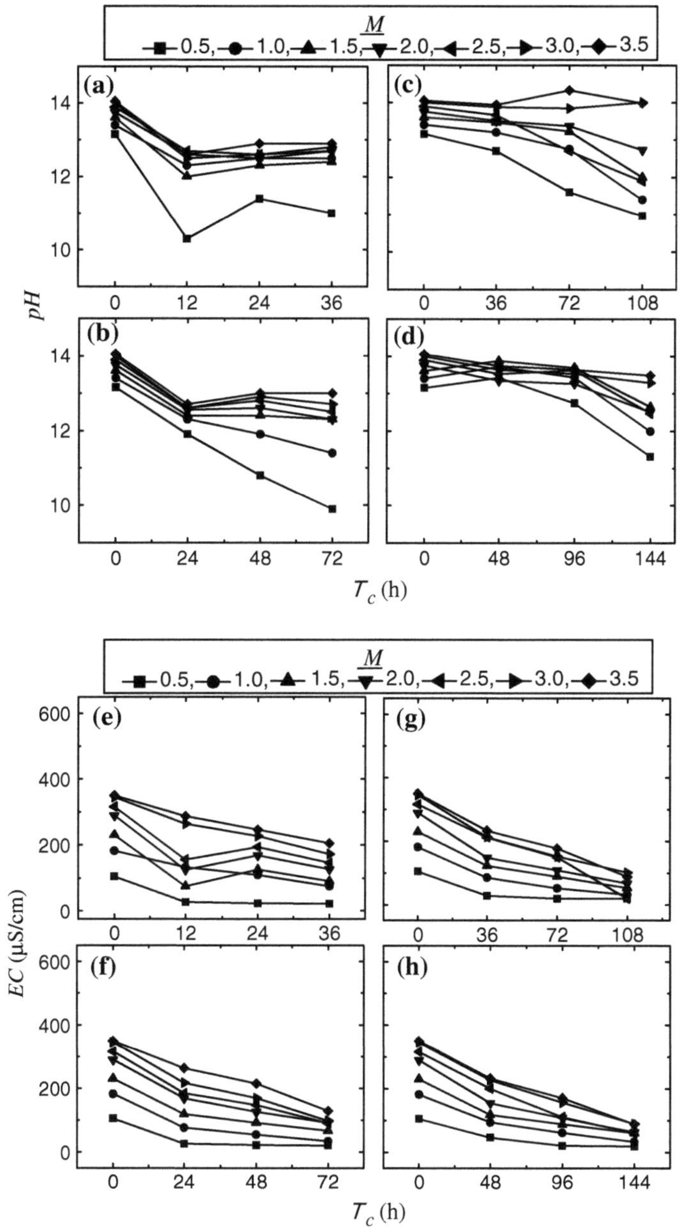

Fig. 6.1 *pH* and *EC* variations of the filtrate with cumulative duration (T_c) for each step duration **a**, **e** 12, **b**, **f** 24, **c**, **g** 36 and **d**, **h** 48 h

increase in the duration and the molarity. This supports the variation in the characteristics of the filtrates, as discussed in Figs. 6.1 and 6.2.

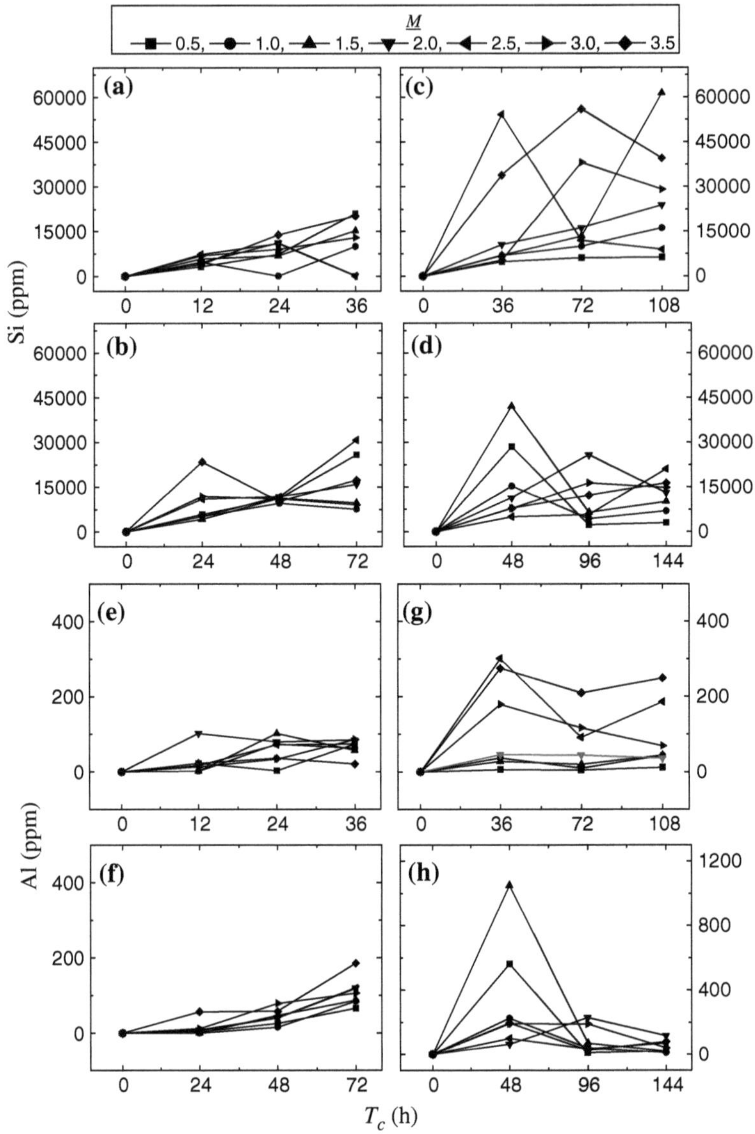

Fig. 6.2 Variation in Si^{4+} and Al^{3+} in the filtrate of the TSA for each step duration **a**, **e** 12, **b**, **f** 24, **c**, **g** 36 and **d**, **h** 48 h

Further, as a result of Step-2 (i.e., R1 and represented by third data point in Fig. 6.3 on time axis), variations in these oxides amidst randomness elucidate significant interaction between the residues and the filtrates, as discussed above. Interestingly, from Fig. 6.3a–h it can be noticed that recycling causes both loss and gain in the major oxides present in the fly ash (refer Fig. 6.3a–h). Such gains are

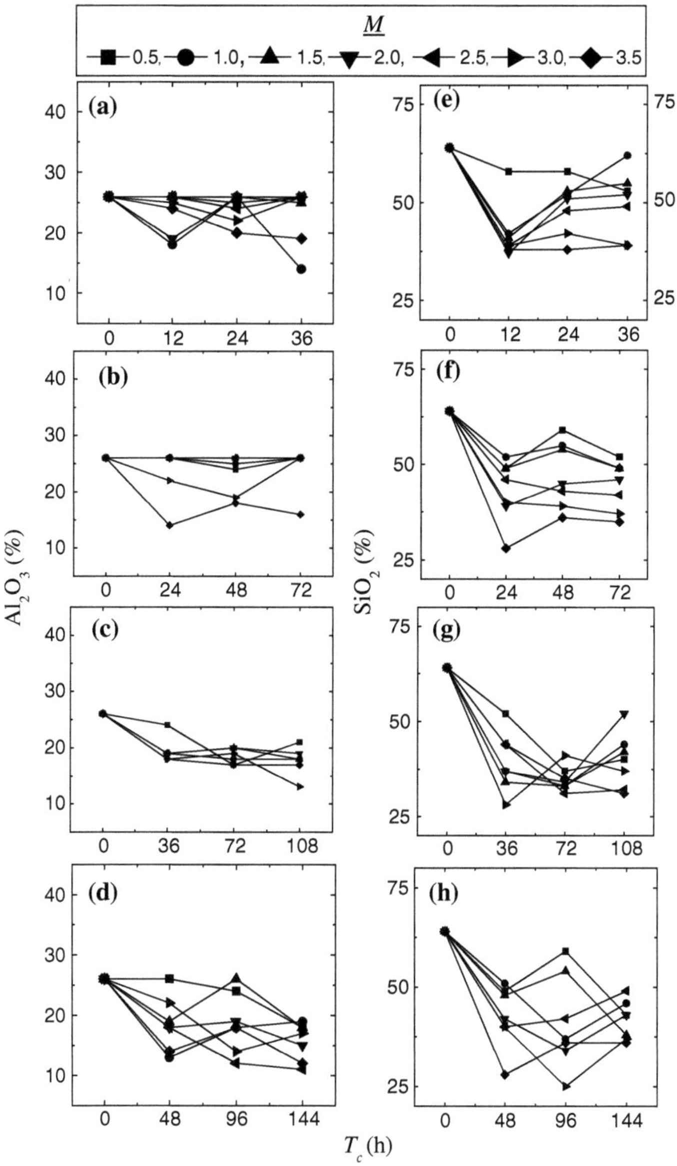

Fig. 6.3 Variations in SiO_2 and Al_2O_3 due to TSA of the fly ash for each step duration **a**, **e** 12, **b**, **f** 24, **c**, **g** 36 and **d**, **h** 48 h

revelation of the gel forms, which get dissociated from the filtrates, are precipitated and get mixed with the residues while the loss demonstrates an increased dissolution of the fly ash residues.

6.1.2.2 Specific Gravity of AAF Obtained from TSA

From Fig. 6.4a–d and Tables 5.12 and 5.13 it can be noticed that there is increase in specific gravity, *G*, due to step-1 (the PT), whereas, only minor variations during the R1 and R2 steps of the treatment. This can have direct correlation with the dissolution of Si and Al species and creation of pores in the particles.

6.1.2.3 Specific Surface Area of AAF Obtained from TSA

From Fig. 6.5a–d and Tables 5.15, 5.16, 5.17 and 5.18 it can be noticed that, *SSA* increases with the increase in step of recycling of the AAF. Incidentally, contrary to all other AAFs, there is significant increase in the *SSA*, in the residue, 1.5-PT-36 (refer Fig. 6.5c). This can be attributed to an intensive etching of the ash particles, leading to its dissolution, which helps the development of micro-sized pores in AAFs. In addition, treatment steps of 36 h exhibits trends, contradictory to other trends. This can be attributed to creation of extra fine particles; which may be amorphous in nature and nano-sized zeolite crystals.

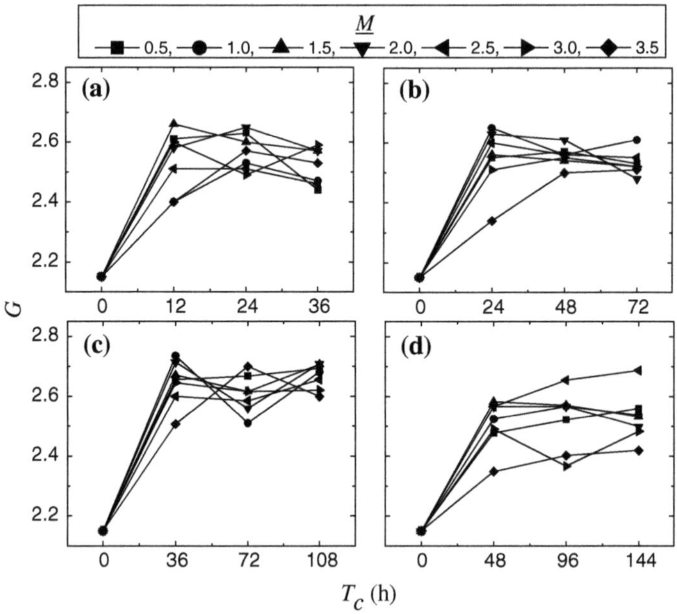

Fig. 6.4 Variation in the specific gravity (*G*) of the AAF with different molarities (*M*) of NaOH solutions and the cumulative durations of the TSA for each activation period **a** 12, **b** 24, **c** 36 and **d** 48 h

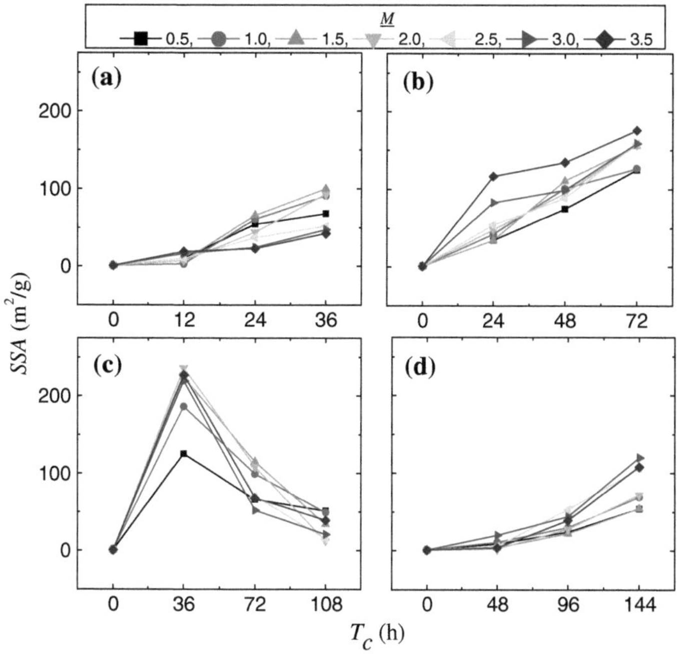

Fig. 6.5 Variation in the specific surface area of the AAF of different steps of the hydrothermal treatment for different activation period in hours **a** 12, **b** 24, **c** 36 and **d** 48 h

6.1.2.4 Cation-Exchange Capacity (*CEC*)

It can be noticed from Fig. 6.6 and Tables 5.25, 5.26, 5.27 and 5.28 that the Step-1 of the TSA results in significant increase in *CEC* of the residues of NaOH solutions of different molarities and activation time. Also, in general, Step-2 (the R1) and Step-3 (the R2) result in an increase in the *CEC* of the fly ash residues. However, a continuous increase in *CEC* can be noticed from the trends depicted in Fig. 7.6b, which demonstrates effectiveness of the R1 and R2. Incidentally, Step-3 of the TSA yields very high *CEC* values (up to 843 meq./100 g) of the residue (designated as 1. 5-R2-24).

6.1.2.5 Mineralogical Characteristics

The residues of the TSA were characterized for their mineralogical characteristics (viz., relative crystallinity (refer Table 5.31), mineral identification and micro-structural alterations). The relative crystallinity of the residues of the Step-1 (i.e., PT and represented by second data point in Fig. 6.7 on time axis) increases, in general, and maximum crystallinity (up to 77 %) corresponds to the residues of the Step-3 (i.e., R2 and represented by fourth data point in Fig. 6.7d on time axis).

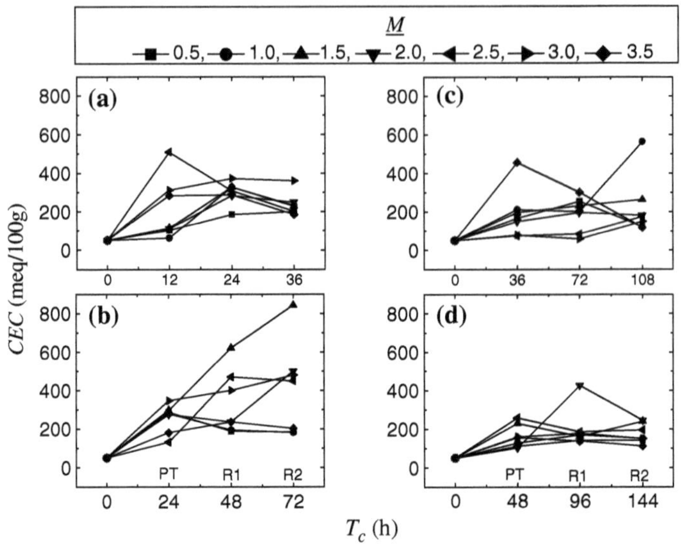

Fig. 6.6 Variation in cation exchange capacity (*CEC*) of the residues due to TSA for durations **a** 12, **b** 24, **c** 36 and **d** 48 h

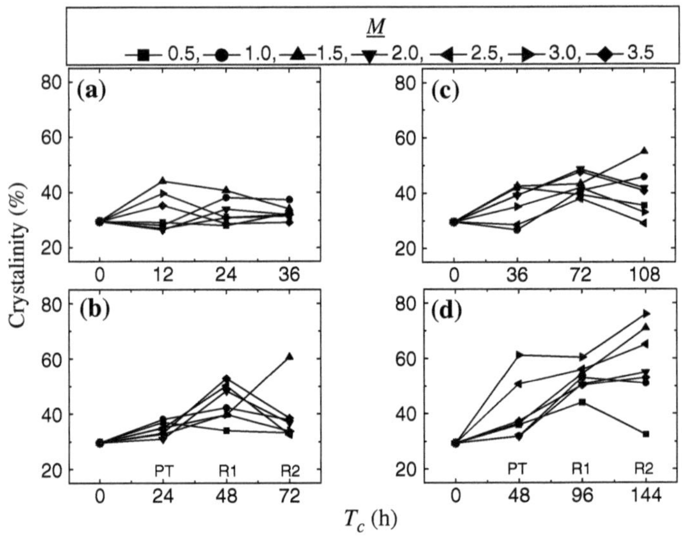

Fig. 6.7 Variation in crystallinity of the residues due to TSA for durations **a** 12, **b** 24, **c** 36 and **d** 48 h

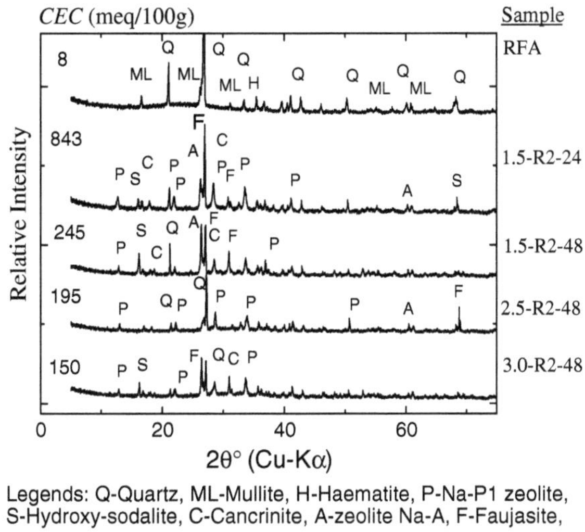

Fig. 6.8 Mineral phase transition in the fly ash and the residues (the marked values of their cation exchange capacity (*CEC*) are in unit, meq./100 g)

The residues, which are significantly crystalline (>60 %, the maximum crystallinity of the residues of Step-1, conforming to those of the conventional methods) were identified as superior residues (i.e., significantly zeolitized residues, SZRs, designated as 1.5-R2-24, 1.5-R2-48, 2.5-R2-48 and 3.0-R2-48). Apart from this, the residues 1.5-R2-24, exhibiting highest *CEC* value (refer Fig. 6.6b), and 3.0-R2-48, exhibiting highest crystallinity (refer Fig. 6.7d) are expected to have significant quantity of superior fly ash zeolites (FAZ). Furthermore, the XRD diffractograms of these residues (refer Fig. 6.8, Tables 5.20, 5.21, 5.22, 5.23 and 6.1) demonstrate a significant difference in chemical and mineralogical characteristics as compared to the original fly ash [4, 5]. It can be noticed from Table 6.1 that the fly ash comprises of mainly Quartz and Mullite (total being ≈42 %), and the rest of it is in amorphous phase. However, the residues at the end of the TSA gain in zeolite Na-P1 (6–8 %, as per Rietveld method and other zeolites, Hydroxy-sodalite, Faujasite, Cancrinite and Na-A as per JCPDS files) [1].

Further, the residue also gets enriched in the amorphous ingredients, as quantified by Rietveld method. The Rietveld result contradicts the findings from the JCPDS [1], which reveals several zeolitic phases in the residues. In such a scenario, the amorphous phase may commensurate with some sodium aluminosilicates (i.e., zeolites, detected by JCPDS) [1] in the residues, which may be associated with low intensity XRD peaks that cannot be detected by the Rietveld software.

A critical analysis of Table 6.2 indicates micro-structural variations in the fly ash and its significantly zeolitized residues (SZR), which can be addressed in terms of different characteristics of the crystallites (viz., size (*CS*), d-spacing of atomic planes, full width at half maximum, FWHM (β), and the h, k, l values of the atomic planes, refer Table 6.2). Incidentally, residues 1.5-R2-24 exhibit less d-spacing (=1.37 Å, corresponding to planar alignment 910 as compared to 220), which corresponds to close atomic packing as compared to other residues.

Table 6.1 Mineral phase identification by JCPDS data files and quantification by Rietveld refinement method

Sample	Phase	Mineral	JCPDS (match)	Rietveld (%)
Fly ash	Crystalline	Quartz	✓	19
		Mullite	✓	23
		Haematite	✓	–
	Amorphous	Glass	–	58
3.0-R2-48	Crystalline	Quartz	✓	14
		Mullite	✓	18
		Na-P1	✓	6
		Faujasite	✓	–
	Amorphous	–	–	62
1.5-R2-24	Crystalline	Quartz	–	12
		Mullite	–	14
		Na-P1	✓	8
		Hydroxy-sodalite	✓	–
		Na-A	✓	–
		Faujasite	✓	–
		Cancrinite	✓	–
	Amorphous	–	–	66

✓ Detected, – not detected

Further, from the data presented in Table 6.2 it can also be derived that *CS*, and planar alignment, *h k l*, corresponding to a specific mineral phase (say zeolite Na-P1 in the residues 1.5-R2-24 and 3.0-R2-48, refer Fig. 6.8) vary and hence crystal anisotropy and structural disorder of the residues, obtained at the end of the TSA, could be verified. In addition, residues 1.5-R2-24 are found to have higher value of the *CS* (=95 nm) as compared to their counterpart 3.0-R2-48, for which *CS* is 60 nm. This can be attributed to better nucleation and crystal growth in residues 1.5-R2-24. Hence, the superiority of residues 1.5-R2-24 gets reconfirmed.

6.1.2.6 Morphological Alterations

It can be noted that spherical particles in the fly ash (refer data points at zero time in Figs. 6.7a, b and 5.10a–k change to polycrystalline shapes in the residues (refer Fig. 6.9c–h). It can also be observed that majority of the spherical particles in the ash are of size 1–50 μm.

The residues 1.5-R2-24 have been found to exhibit majority of newly formed spherules of diameter ranging from 30 to 50 nm, which commensurate with the zeolite, Na-P1 (designated by P; the fibrous lumps in Fig. 6.9c), which is an improvement over the residues formed by resorting to conventional methods [2, 6, 7]. In addition, fewer ball shaped crystals viz. Hydroxysodalite [7], Faujasite [8],

Table 6.2 Analysis of the X-ray diffractograms of the original fly ash and superior residues

2θ (°)	β (°)	*CS* (nm)	Sample	*d* (Å)	*I* (%)	Mineral phase	Alignment (*h*, *k*, *l*)
12.79	0.23	34	1.5-R2-24	6.92	14	Na-P1	220
12.92	0.20	40	3.0-R2-48	6.85	14	Na-P1	110
16.10	0.16	48	1.5-R2-24	5.5	10	Hydroxy-sodalite	210
16.31	0.06	120	3.0-R2-48	5.43	33	Mullite	211
16.66	0.20	40	1.5-R2-24	5.32	9	Cancrinite	011
21.04	0.20	40	Fly ash	4.22	23	Quartz	310
21.34	0.13	60	3.0-R2-48	4.16	16	Na-P1	220
21.19	0.16	48	1.5-R2-24	4.19	25	Na-P1	211
22.02	0.26	30	1.5-R2-24	4.03	14	Na-P1	131
26.21	0.15	54	1.5-R2-24	3.39	35	Faujasite	400
26.43	0.11	69	3.0-R2-48	3.37	100	Faujasite, Quartz	600
26.53	0.10	81	1.5-R2-24	3.35	26	Faujasite	420
26.82	0.20	40	Fly ash	3.32	100	Quartz	333
26.93	0.13	61	1.5-R2-24	3.31	100	Na-P1	642
27.13	0.10	81	3.0-R2-48	3.28	94	Quartz	323
28.42	0.10	81	1.5-R2-24	3.13	32	Cancrinite, Na-P1	301
28.61	0.20	41	3.0-R2-48	3.12	36	Na-P1	115
30.83	0.16	49	1.5-R2-24	2.9	14	Faujasite	411
30.99	0.08	98	3.0-R2-48	2.88	51	Faujasite	751
33.54	0.20	41	1.5-R2-24	2.67	27	Na-P1	312
33.70	0.16	49	3.0-R2-48	2.65	49	Faujasite	312
35.68	0.13	62	3.0-R2-48	2.51	26	Mullite	020
41.30	0.16	50	3.0-R2-48	2.18	24	Mullite	080
50.40	0.23	37	1.5-R2-24	1.81	16	Na-P1	416
68.43	0.10	95	1.5-R2-24	1.37	16	Na-P1	910

Note CS crystallite size (computed by employing Scherrer's formula), β the FWHM, θ diffraction angle, *d* inter-planar spacing, *I* relative intensity of the peaks

Cancrinite [9] and zeolite Na-A [10], marked as S, F, C, and A, respectively, can also be observed in Fig. 6.9c.

On the other hand, from Fig. 6.9e it can be observed that the residues 3.0-R2-48 also contain the ball shaped crystals viz. Hydroxy-sodalite, larger in size ranging from 900 to 1000 nm. Apart from this, fewer crystal shapes and sizes of Faujasite (prismatically shaped, 1 µm long) and zeolite Na-P1 (spherules of 100 nm size) are also found to be present in these residues. Accordingly, the higher *CEC* of the residues 1.5-R2-24, being the most superior residues, can be attributed to its polycrystalline zeolitic phase (viz., Na-P1, Cancrinite and Na-A) as compared to the residues 3.0-R2-48, which comprises of different types of zeolites (viz., Hydroxy-sodalite and Faujasite) and hence exhibits a lesser *CEC*.

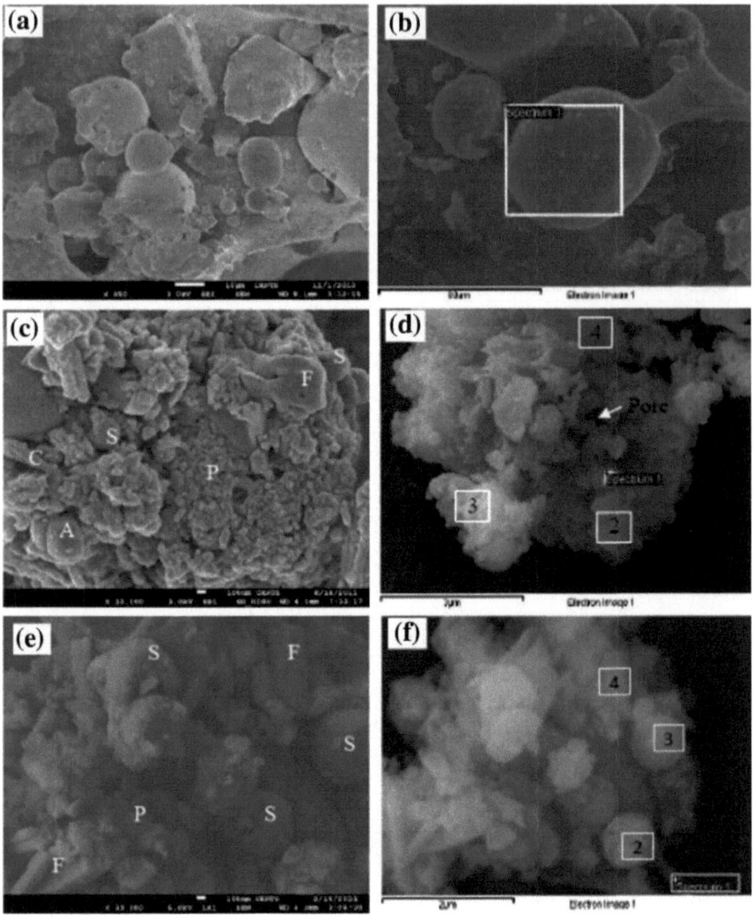

Fig. 6.9 FEG-SEM/EDS micrographs **a**, **b** Fly ash. **c**, **d** Residues, 1.5-R2-24 and **e**, **f** Residues, 3.0-R2-48, where numbers marked *within the boxes* represent the EDS spectrum number

6.1.2.7 Surface Elemental Heterogeneity

From Table 6.3 and Fig. 5.10a–k i.e. the energy dispersive spectrometry, EDS, results it can be observed that the residues 3.0-R2-48 comprise of fewer Si and Al atoms and more of Na and oxygen atoms as compared to the residues 1.5-R2-24 (refer Table 6.4). However, more number of sodium and oxygen atoms in the residues 3.0-R2-48 indicate their polymerization and hence, higher crystallinity of the residues (refer Fig. 6.9d) gets verified. By conducting a reanalysis of the data presented in Table 6.3, the two ratios, Na/Al and Si/Al, were computed and their values are listed in Table 6.5.

It is apparent from Table 6.5 that the value of Na/Al for the residues 1.5-R2-24 varies from 0.477 to 0.783, which is <1 and hence is conforming to early phase of

Table 6.3 Elemental composition on the crystal surface in the hopper fly ash and significantly zeolitized residues (SZRs)

Elements	% by weight on particle surface within the selected EDS spectrum (shown by number in parenthesis)								
	Fly ash	Residue 1.5-R2-24				Residue 3.0-R2-48			
	(1)	(1)	(2)	(3)	(4)	(1)	(2)	(3)	(4)
O	50.50	49.81	50.01	51.92	34.53	57.42	55.88	56.1	53.07
Al	14.62	8.84	10.28	10.21	10.62	3.85	10.59	9.51	7.94
Si	23.24	23.57	29.44	18.48	29.50	9.74	19.73	17.46	14.74
Fe	1.08	–	2.56	1.30	–	–	0.72	4.83	0.47
Na	–	5.66	7.0	7.99	5.07	28.99	12.82	10.97	17.45
K	0.49	0.33	–	–	0.78	–	0.26	–	–
Ca	1.55	0.55	0.71	1.01	2.48	–	–	–	0.26
Ti	1.650	0.40	–	–	–	–	–	–	–

– Below detection limit

Table 6.4 EDS data for spectrums 1–4 of the residue 1.5-R2-24 (refer Fig. 5.10o–i, ii)

Spectrum 1			Spectrum 2		
Element	Weight (%)	Atomic (%)	Element	Weight (%)	Atomic (%)
O	49.89	67.26	O	50.01	63.49
Na	5.66	5.31	Na	7.00	6.18
Al	8.84	7.07	Al	10.28	7.74
Si	23.57	18.11	Si	29.44	21.29
K	0.33	0.18	Ca	0.71	0.36
Ca	0.55	0.30	Fe	2.56	0.93
Ti	0.40	0.18	Cu	0.00	0.00
Fe	1.30	0.50	Zn	0.00	0.00
Cu	0.00	0.00	Bi	0.00	0.00
Re	9.46	1.10			
Spectrum 3			Spectrum 4		
O	51.92	68.43	O	34.53	52.90
Na	7.99	7.33	Na	5.07	5.40
Al	10.21	7.97	Al	10.62	9.65
Si	18.48	13.87	Si	29.50	25.75
Ca	1.01	0.53	K	0.78	0.49
Ti	1.05	0.46	Ca	2.48	1.52
Fe	1.30	0.49	Ti	1.25	0.64
Cu	0.00	0.00	Fe	5.15	2.26
Re	8.04	0.91	Cu	0.00	0.00
			Re	10.62	1.40
			Bi	0.00	0.00

Table 6.5 Important parameters (in ratio form) of the fly ash and the superior residues

Ratio	Computed value of the ratio from selected EDS spectrum (shown by number in parenthesis)								
	Flyash	Residue 1.5-R2-24				Residue 3.0-R2-48			
	(1)	(1)	(2)	(3)	(4)	(1)	(2)	(3)	(4)
Na/Al	0	0.64	0.68	0.78	0.47	7.53	1.21	1.15	2.19
Si/Al	1.59	2.66	2.86	1.81	2.77	2.53	1.86	1.83	1.85

zeolites viz. Na/Al = 1, corresponding to chemical formula $Na_6Al_6Si_{10}O_{32} \cdot 12H_2O$ for the zeolite Na-P1 [11, 12]. This can be attributed to an increased alumination (i.e., incorporation of Al) of Si tetrahedron, which could be responsible for higher *CEC* and hence better grade of the FAZ in the residues. On the other hand, variation of Na/Al and Si/Al from 1.151 to 7.53 and 1.85 to 2.77, respectively, in case of residues 3.0-RC2-48 could be ascribed to the presence of Hydroxy-sodalite ($Na_8(AlSiO_4)_6 \cdot (OH)_2 \cdot 3H_2O$ and designated by S) in majority and Faujasite ($Na_{58}Al_{58}Si_{134}O_{384}$ and designated by F) in traces [1, 10], as depicted in Fig. 6.9. However, the presence of these zeolites in residues 3.0-R2-48 is responsible for their lower *CEC*, as depicted in Fig. 6.6d. These observations hint at improved zeolitization of the fly ash compared to data reported in the literature [13].

6.1.2.8 Surface Pores

Based on reanalysis of the FEG-SEM micrographs (refer Figs. 7.9a and 5.10a–k) the surface pores have been magnified and clearly exhibited in Fig. 6.10. It can be noticed from Fig. 6.9 that the spherical particles of the original fly ash have smooth

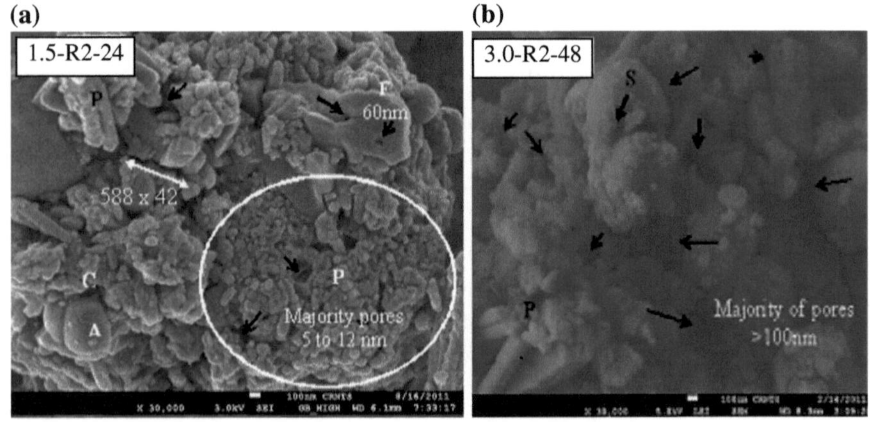

Fig. 6.10 Various sizes of surface pores in the residues **a** 1.5-R2-24 and **b** 3.0-R2-48, where *A*, *C*, *F*, *S* and *P* are designations for fly ash zeolites: Na-A, Cancrinite, Faujasite, Hydroxy-sodalite and Na-P1, respectively

boundary and no pores on their external surface. On the other hand, several open pores could be noticed on the surface of the residues (refer micrographs of the residues 1.5-R2-24, exhibited in Fig. 6.10).

With the help of ImageJ software [14], it can be inferred that some pores (i.e., shown by black arrows in Fig. 6.10) are ranging from 155 to 588 nm in major dimension and majority of pores exhibit size in the range of 5–12 nm (marked by the circled zone on the micrograph in Fig. 6.10a), and hence it can be concluded that the TSA of the fly ash results in smaller pores. This can be major attribute for high *CEC* of the residues, 1.5-R2-24. On the contrary, residues 3.0-R2-48 (refer Fig. 6.10) highlights major dimensions of open surface pores (i.e., shown by black arrows), where most of the pores have sizes more than 100 nm (i.e., higher than those in the residues 1.5-R2-24), and hence results in lower *CEC* than its counterpart.

6.1.2.9 Comparison of the Cost of the Residues Obtained from TSA and Conventional Methods

In order to compare the costs involved in the TSA method and other conventional methods (viz., one step and two steps) for synthesis of the ash zeolites, various parameters (viz., type of the reagent used, duration of the steps, cost of the equipment and energy consumed) were considered and the details are presented in Table 6.6.

It should be noted here that the one- and two- step methods mostly employ a closed reflux system [15], which facilitate hydrothermal activation of the fly ash at elevated pressure but are expensive. In addition, the two-step method employs chemicals like NaOH and $NaAlO_2$ [15], which also add to the overall cost of synthesis of the fly ash zeolites. On the contrary, the TSA being conducted in an open reflux system (refer Fig. 5.1), the cost of synthesis of zeolites is reasonably low. Also, lower energy consumption ($\approx$72 kWh) supports the superiority of the TSA, as compared to the two-step method. In terms of purity of the zeolites, the process adopted in the TSA (refer Fig. 5.2) results in enhanced cation-exchange capacity, *CEC*, of the residues of the fly ah (843 meq./100 g) as compared to the conventional methods (388 meq./100 g for 1-step activation and 250 meq./100 g for 2-step activation) [15]. Apart from this, the multiple recycling of the filtrates before their final disposal, in the TSA, is helpful in reducing the *pH* and concentration of the heavy metal ions (viz., Si and Al) present in them, which is not the case with the conventional methods.

Table 6.6 Cost comparison of three hydrothermal processes of zeolite synthesis from fly ash

Details	TSA (Jha and Singh) [17]				2-Step activation (Hollman et al.) [15]			1-Step activation (Hollman et al.) [15]		
	Step	Unit (kWh)	Rate#	Amount#	Unit (kWh)	Rate#	Amount#	Unit (kWh)	Rate#	Amount#
Cost of electricity, 1000 W heater for 24 h treatment	PT	24	0.11	2.64	6	0.11	0.66	96	0.11	10.56
	R1	24	0.11	2.64	48	0.11	5.28	*		
	R2	24	0.11	2.64	*					
Cost of fly ash	PT	–	0.02 (≈1 kg)	0.002 (≈100 g)	0.002 (≈100 g)	*			0.02 (≈1 kg)	0.002 (≈100 g)
Cost of NaOH	PT	–	2.78 (≈1 kg)	0.16 (≈60 g)	*	2.78 (≈1 kg)	0.05 (≈20 g)	*	2.78 (≈1 kg)	0.05 (≈20 g)
Cost of $NaAlO_2$	R1	*				2.78 (≈1 kg)	0.05 (≈20 g)	*		
Open reflux set up	PT	*	55.57	55.57	*					
Closed reflux set up	*				*		370.5	*		370.5
Initial cost of experimental set up				55.57			370.5			370.5
Recurring experimental cost				8.082			6.06			10.61

*Data not available

#Rate and amounts are in US dollar

6.2 Purification of Fly Ash Zeolites

Over the decades, fly ash zeolites have been synthesized from fly ash by resorting to its hydrothermal activation with NaOH, in one step. However, the activated ash (the residue) has been found to exhibit lower *CEC*. Based on X-ray diffraction spectrometry, Quartz and Mullite (i.e., the inactivated or residual fly ash), are found to be present in the residues as impurities in the zeolites (i.e., Na-P1, in majority). Also, such impurities have been verified from the field emission gun-scanning electron microscopy (FEG-SEM) of the residue. The purification of these zeolites is still a big challenge [16]. In such a situation, "*three-step hydrothermal process*" also ascertains significant gains in their characteristics viz. *CEC* by 185 and 14 % (refer Fig. 6.11) and specific surface area by 478 and 33.42 % (refer Fig. 6.12b), of the residues obtained after Step-2 and Step-3 of this process, respectively. In addition, lowering of the specific gravity and improvement in the SiO_2/Al_2O_3 ratio of the residue are indicative of the formation of more porous and silica rich zeolites, in a purified form. Accordingly, an attempt has been made to assimilate all outcomes of the adopted process and develop a conceptual model to exhibit the complexities involved in overall conversion of the fly ash to pure zeolites [17–19].

6.2.1 *Monitoring Relative Variations Between Two Steps of TSA*

From Fig. 6.11a, b it is apparent that Step-1 (the PT) causes an increase in the *G* and *SSA* of the residues by 13 and 84 %, respectively, corresponding to 1.5 M

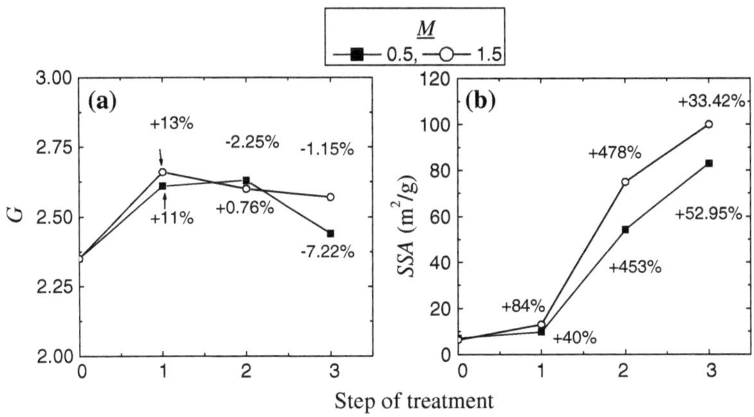

Fig. 6.11 Effects of recycling and step of the treatment on **a** *G* and **b** *SSA* of the fly ash and residues, where percentage increase and decrease in the parameter, due to each step, are marked by the *plus* and *minus sign*, respectively

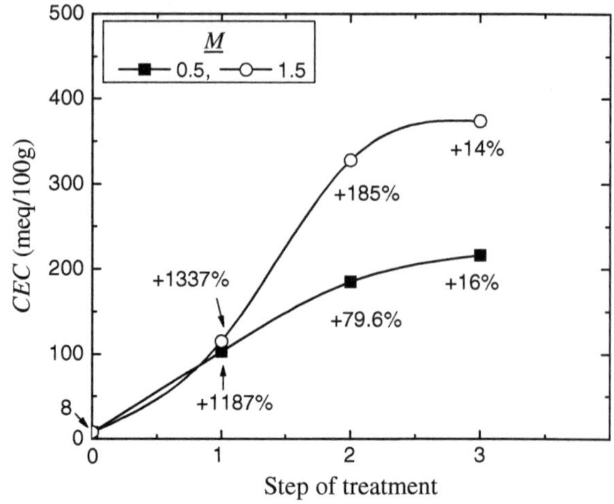

Fig. 6.12 Variation in the *CEC* of the samples, where percentage increase in the *CEC*, after each step has been marked, by a *plus sign*

NaOH, which can be opined to be a superior reagent than 0.5 M NaOH. These outcomes are in full agreement to Kolay and Singh [20] and can be attributed to various effects of ash-alkali interaction in Step-1 (the PT) such as; (i) etching at fewer areas on interacting surface of the ash and (ii) creation of surface pores, visible in the micrographs, as discussed in the following section.

The pores may also result in release of entrapped gases from inner region of the ash particles and hence higher value of specific gravity (*G*) of the residues. Conversely, Step-2 (the R1) causes a minor change in the specific gravity, while a significant increase (by 478 %) in the *SSA* is noticeable. This could be due to significant dissolution of the impurities and precipitation of the resulting soluble Si, Al and Na as smaller particles in size up to 50 nm, which could get incorporated in the residues of Step-2. Further, Step-3 results in reduction of *G* and continuous increase of the *SSA*.

Incidentally, a decrease in *G* of the residues may be attributed to: (i) enhanced formation of porous compounds and (ii) polymerization of Si and Al tetrahedra, which also corresponds to increased volume of a zeolitic framework structure, present in the residues. Based on variations of *G* and *SSA* it can be inferred that Steps-2 and 3 (the R2), are indicative of better zeolitic characteristics of the residues (i.e., purified form of residues of Step-1).

Also from Tables 5.20, 5.21, 5.22 and 5.23 reduction in most of the oxides (viz., SiO_2, Fe_2O_3, BaO, P_2O_5 and TiO_2) after Step-1 gets verified. Especially, reduced quantity of heavy metal oxides is another controlling factor for reduced *G* values of the residues after Steps-2 and 3, corresponding to 1.5 M NaOH. In addition, from Table 6.7 it can be noticed that residue of Step-1 (the PT) and 1.5 M NaOH exhibits SiO_2/Al_2O_3 ratio i.e. 58 % lower than that of the original fly ash, which indicates enhanced dissolution of SiO_2 (Quartz).

Table 6.7 Important parameters of the fly ash, RFA and the residues, AAF

Ratio	RFA	AAF					
		0.5-PT-12	0.5-R1-12	0.5-R2-12	1.5-PT-12	1.5-R1-12	1.5-R2-12
SiO_2/Al_2O_3	2.45	2.07 (−16)	2.28 (+10)	1.55 (−32)	1.03 (−58)	2.01 (+95)	2.20 (+9)
Na_2O/Al_2O_3	0.002	0.16 (+7900)	0.21 (+31)	0.17 (−19)	0.23 (+11400)	0.38 (+65)	0.37 (−3)

Note A number under parenthesis represents percentage increase or decrease, as designated by + or − sign, respectively due to its stage in comparison to immediate previous stage of the treatment

Incidentally, a significant increase in Na_2O/Al_2O_3 has been observed in residues obtained from 1.5 M NaOH activation (11,400 % for 1.5-PT-12, refer Table 6.7) as compared to 0.5 M NaOH activation (7900 % for 0.5-PT-12, refer Table 6.7). This indicates incorporation of more Na^+ cations in 1.5-PT-12 residue and hence it's improved *CEC* by 1337 % (refer Fig. 6.12), as compared to the fly ash. This may be due to significant reduction in Si content and/or increase in Al content (refer Table 6.7), which gets reflected from reduction of XRD peak intensity of Quartz and Mullite, present as impurities in the zeolite Na-P1 (refer Fig. 6.13).

Based on Fig. 6.13, a summary of the mineralogical phase transition in the residues obtained from TSA, has been listed in Table 6.8. The presence of minerals in the residues could be identified on the basis of their XRD peak intensity (viz., higher peak intensity indicates higher mineral content and low peak intensity indicates lower mineral content). From this table, the increase in amount of fly ash zeolites (viz., Na-P1 and Faujasite) and decrease in that of Quartz and Mullite, in the residues of Step-2 (corresponding to molarity M = 1.5), establish their superiority over other residues (M = 0.5). These zeolites also get verified from Fig. 6.14, which exhibits several morphologies such as; (i) radiating fibers and rod shaped morphology resembling zeolites Na-P1 length ranging from 450 to 800 nm

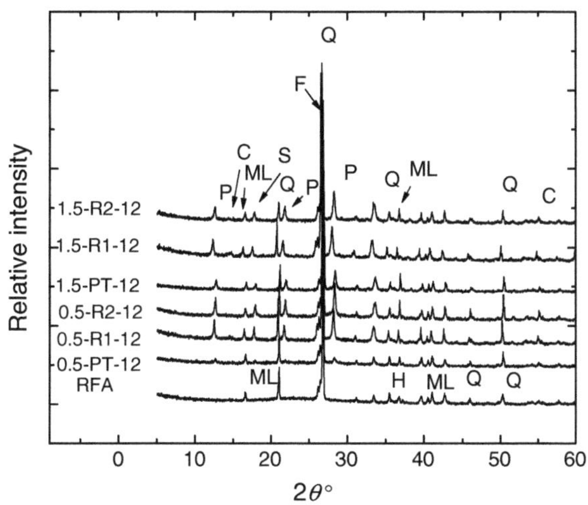

Fig. 6.13 Variation in XRD patterns of the samples, where the residues correspond to three steps (viz., PT, R1 and R2) of treatment, each for activation time, 12 h

Table 6.8 Minerals present in the RFA and AAF samples

Mineral	RFA	0.5-PT-12	0.5-R1-12	0.5-R2-12	1.5-PT-12	1.5-R1-12	1.5-R2-12
Quartz	+++	++	–	–	++	–	–
Mullite	+++	++	+	+	++	+	–
Na-P1	–	+	+++	++	++	+++	++
Na-X (Faujasite)	–	–	+	++	+	+++	++
Chabazite	–	–	–	–	–	+	+
Hydroxysodalite	–	–	+	+		+	+

Note +++, ++, +, – Represent very high, high and low content, respectively, whereas, dash represents not identifiable

(refer Fig. 6.14b, c) and (ii) ball shaped morphology in diameter ranging from 75 to 100 nm, which resembles to Hydroxysodalite (refer Fig. 6.14d). In addition, Fig. 6.14f exhibits honeycombs (i.e., agglomerates of premature zeolites i.e., under process of precipitation transforming into sheet shaped Chabazite). These alterations commensurate with increasing inter particle pore size, as noticed in the micrographs (refer Fig. 6.14b–g). Based on these observations, it can be inferred that three-step activation is more effective than one step (viz. conventional) treatment of the fly ash.

In addition, Fig. 6.14b–d exhibits remarkable growth in width of these zeolitic shapes, which increase from 80 to 100 nm. These could be responsible to new peaks representing zeolites in the residues of Step-2 (refer Fig. 6.13) from 1.5 M NaOH solution, which causes increase in Na_2O/Al_2O_3 and SiO_2/Al_2O ratios by 65 and 95 %, respectively (refer Table 6.7).

Interestingly, it can be noticed from Fig. 6.12 that, *CEC* of these samples increases (by 1187–1337 %), due to Step-1. The Step-2 (the R1) causes further improvement (by 185 %) in the *CEC*, for 1.5 M NaOH. However, Step-3 (the R2) results in a minor increase in the *CEC*. Thus, Step-2 and 1.5 M NaOH can be thought to be a better combination, which results in a larger gain in the *CEC* of the residues. This can be attributed to an augmented formation of a new zeolite (ball shaped Hydroxysodalite) along with Na-P1 and hence, improvement in their purity. In agreement to Grutzeck and Siemer [21], this can be attributed to increased linkage of cations (Na^+), and hence more *CEC* by 105.4 % (refer Fig. 6.12). Conversely, Step-3 causes significant decrease in Na_2O/Al_2O_3 and SiO_2/Al_2O in case of residues from 0.5 M NaOH. This can be ascribed to conversion of Na-P1 to Hydroxy-sodalite from Step-2 to Step-3 (refer Figs. 6.13 and 6.14), and this conforms to less increase in the *CEC* (refer Fig. 6.12).

Furthermore, Fig. 6.14 exhibits significant transitions in shape and size of the crystalline constituents from the RFA to the residues of Steps-1, 2 and 3 of the treatments. In addition, the particle size (=2.5–30 μm) of the minerals in the fly ash reduces to 50–800 nm in the residues, depending on the type of alkali solution and the step of the treatment. From the micrographs it can be inferred that the shape of the newly formed crystals in the residues vary from rod shaped (marked as P), in the

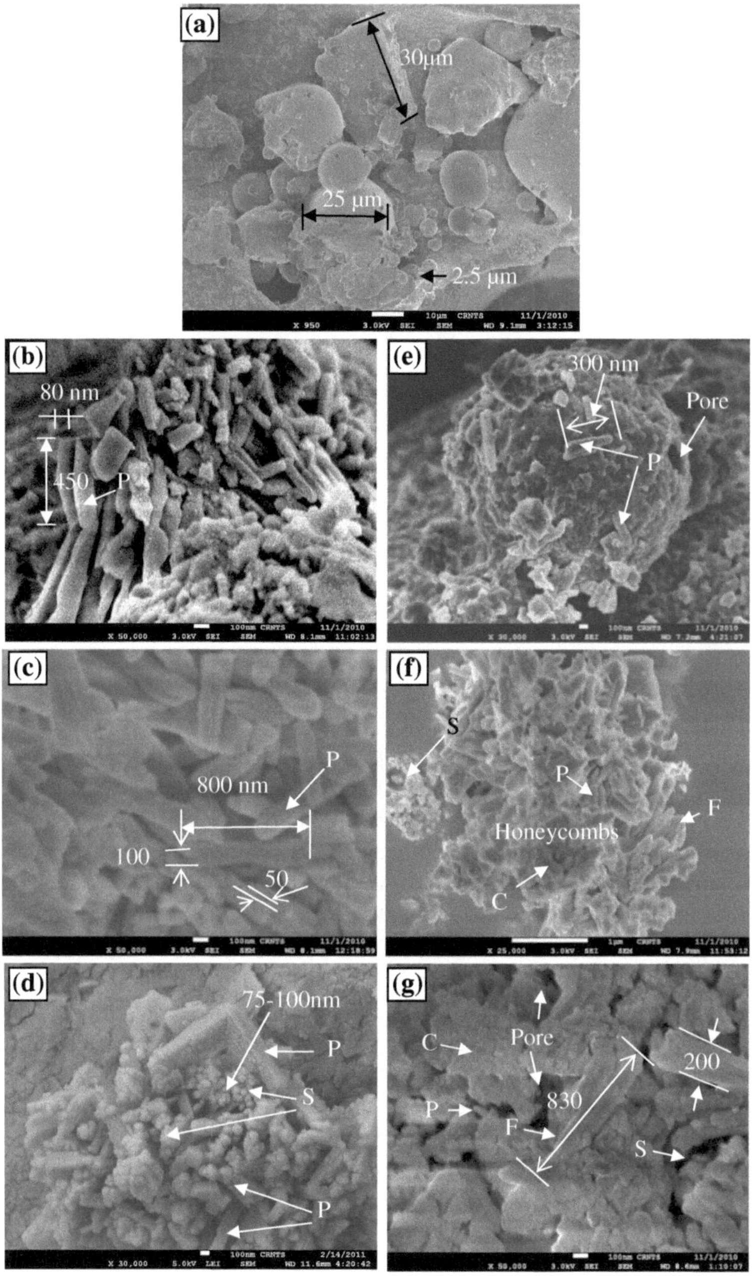

Fig. 6.14 FEG-SEM micrographs of samples **a** RFA, **b** 0.5-PT-12, **c** 0.5-R1-12, **d** 0.5-R2-12, **e** 1.5-PT-12, **f** 1.5-R1-12 and **g** 1.5-R2-12, where *P*, *C*, *S* and *F* designate the zeolites Na-P1, Chabazite, Hydroxy-sodalite and Faujasite, respectively

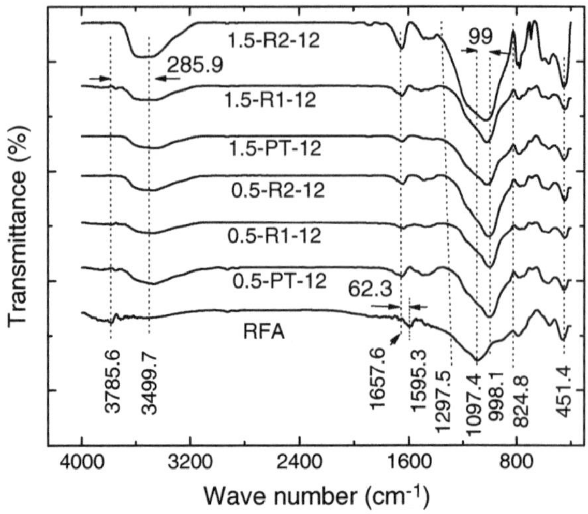

Fig. 6.15 FT-IR spectra of the RFA and the residues in the frequency range 4000–400 cm^{-1}, where all the residues correspond to activation time of 12 h

Step-1, to spheroidal balls (marked as S) and/or honeycombs (C) in the recycling and Step-3 of the treatments.

The mineralogical phase transition in the residues of the three steps (refer Table 6.8) gets verified from the FT-IR spectra and TG-DTA curves presented in Figs. 7.15 and 7.16, respectively. The disappearance of a weak band (i.e. Si-OH, silanol) at high frequency (3785.6 cm^{-1}) from the FT-IR spectrum of the fly ash (refer Fig. 6.15), and simultaneously appearance of broader bands (i.e. Si-OH-Al) at low frequency (3499 cm^{-1}) in that of the residues confirm significant conversion of Si and Al tetrahedra from Quartz and Mullite to new product, which comprises of new structural bonds. Such development favors mutual coordination between Si and Al tetrahedra by a bridging hydroxyl functional group in the residues.

This is in agreement with the findings reported by the previous researchers [13, 22]. However, a sharper band at 1657.6 cm^{-1} confirms relatively more adsorbed water in the residues 1.5-R2-12, which indicates (i) their higher surface negativity and (ii) increased presence of $[AlO_4]^{5-}$ tetrahedron in the zeolitized residues. This again gets substantiated from another sharper and deeper band at 998.1 cm^{-1}, which corresponds to higher crystallinity of the zeolitic framework [Si-O-Al] bonds in these residues 1.5-R2-12. In general, the intensity of these bands increases at the shifted low frequency in the spectra of Step-2 (the R1) and Step-3 (the R2).

These findings are indicative of increased stretching vibration of Si-O-Al bonds, which reveal higher yield of fly ash zeolites in the residues of these steps of the treatment [13, 22–27]. These are responsible for major structural alteration in the residues and growth of more distinct morphology of the crystals, which resemble to common fly ash zeolites, as marked in the micrographs (refer Fig. 6.14).

Apart from these alterations, increase in intensity at frequencies 824.8 and 451.4 cm^{-1} can be attributed to bending vibrations, which could be due to improved porosity of the residues, as reported by Jacob and Mortier [23]. This gets confirmed by increased weight loss (refer Fig. 6.16c) and a distinct endothermic peak (refer Fig. 6.16d), which hints at increased presence of surface and structural water in the residues 1.5-R2-12, as compared to its counterpart 0.5-R2-12, which undergoes sudden weight loss. This could be because of carbonaceous impurities in the residues and their conversion to gases like CO_2 [13], which are not in case of the residues corresponding to 1.5 M NaOH (1.5-R2-12). In addition, lowering of DTA curve (refer Fig. 6.16d) of residues 1.5-R2-12, as compared to 1.5-R1-12 and 1.5-PT-12 is also an indication of less heat flow capacity of such residues and hence less mineral phase transition. Incidentally, residues 1.5-R2-12 exhibit many TGA peaks beyond 1000 °C (refer Fig. 6.16c), which can be attributed to thermal conversion of some minerals to gaseous products entrapped in the zeolitic pores. However, a flatter DTA curve reveals (i) negligible variation in mineralogy of the residues; (ii) increased thermal stability of the residues obtained from Step-3 as compared to Steps-1 and 2 and hence (iii) more purified form of zeolites from Step-3 treatment with 1.5 M NaOH and activation time for 24 h.

6.2.2 *A Conceptual Model for Step Wise Purification of FAZ*

The improvement in various characteristics discussed above has been referred to develop a model for the transitions, due to three-step processes and to represent them in terms of physical transformation, from a residue to a fly ash zeolite. A conceptual model of a fly ash particle and the process of three-step synthesis of zeolites have been presented in Figs. 6.17 and 6.18, respectively. As reported by Inada et al. [28], an actual fly ash particle, as depicted in Fig. 6.17, has been opined to contain glass, in the outer region, and this interacts first with the alkali during Step-1 of the treatment and gets etched [16, 19]. Further, underlying Quartz interacts with the supernatant during the Step-2. Subsequently, as reported by Brouwers et al. [29], the inner most Mullite could get exposed and activated up to Step-3. Based on such activation of the fly ash particle, remarkable increase in *CEC* is observed (refer Fig. 6.12), with an increase in the recycling from Step-2 to Step-3. This highlights significant reduction in the impurities (IMP) in the residues, which gets confirmed from enhanced FT-IR band intensity at frequency 3499.7 cm^{-1} (which corresponds to Si-OH-Al bond, refer Fig. 6.15) leading to high electronegativity of the residues. In addition, high yield of the fly ash zeolites can also be confirmed from XRD diffractograms and FEG-SEM micrographs of the residues (refer Figs. 6.13 and 6.14).

As zeolitization of fly ash is a complex phenomenon, Fig. 6.18 intends to simplify complex effects of successive activations of the fly ash, in three-steps. In such a situation, grinding of a residue, may cause peeling off and crumbling the deposited crystals of the FAZ (refer Fig. 6.14) and thus exposes the roughened boundary of impurities, IMP from Step-1 and Step-2. In fact, the activated boundary contains many nano-sized pores, as observed in Fig. 6.14b–g, which

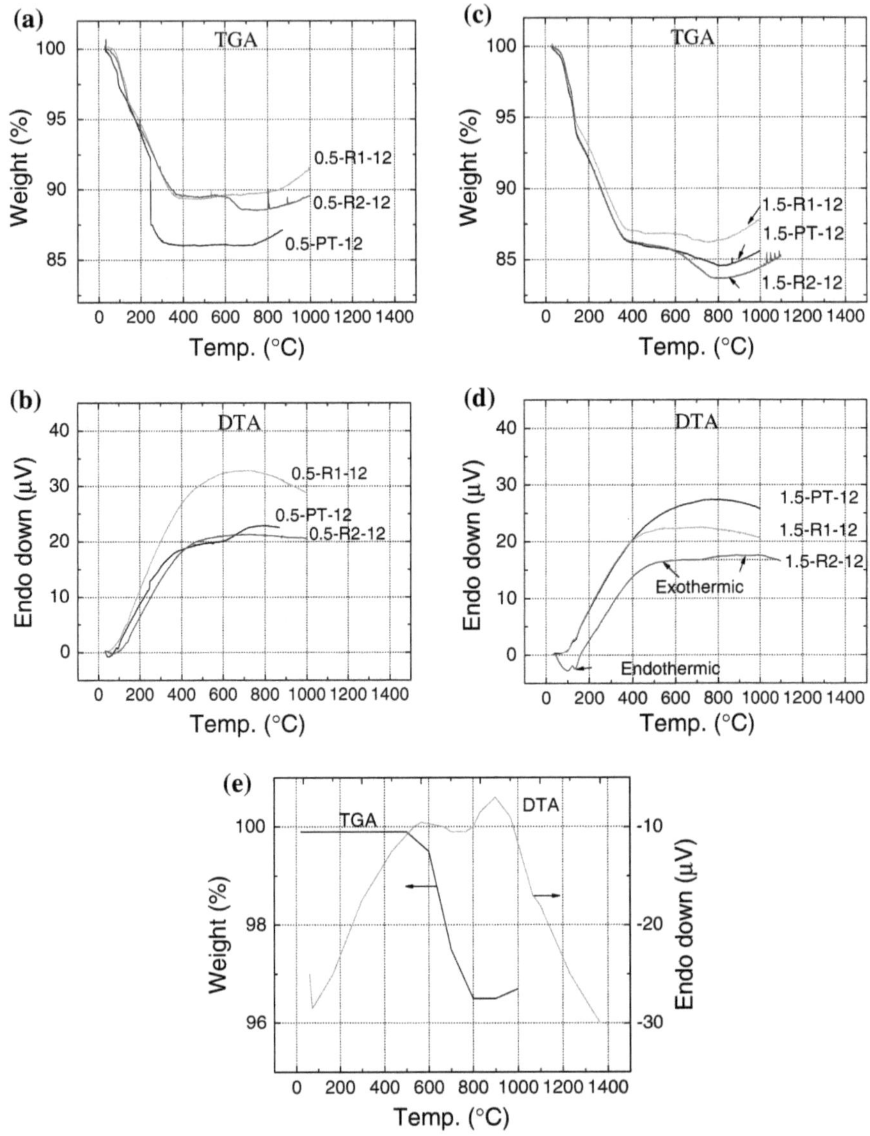

Fig. 6.16 TG-DTA analysis of residues of three steps of hydrothermal treatment with **a**, **b** 0.5 M NaOH, **c**, **d** 1.5 M NaOH and **e** Fly ash, where PT, R1 and R2 are designations for Steps-1, 2 and 3, respectively, for activation time of 12 h

facilitates increased surface contact (refer Fig. 6.11b) and also improved interaction through a pore, up to inner region of the RFA particle, with the supernatant. Accordingly, residues of the Steps-2 and 3 have been thought to contain a number of thermally stable shapes (shown by white color in Fig. 6.18), which commensurate to conversion of primarily synthesized rod shaped zeolite Na-P1 to

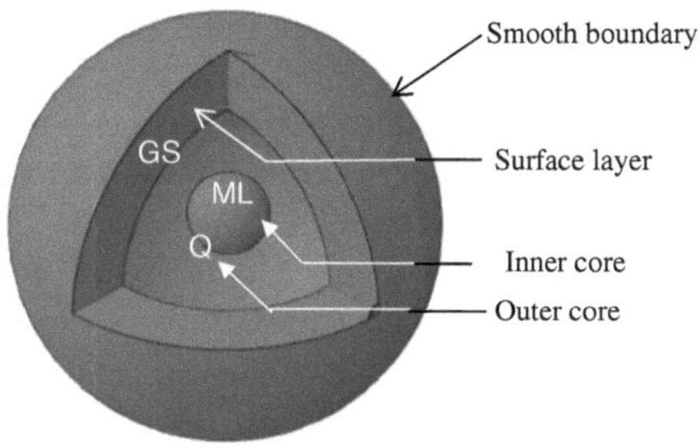

Fig. 6.17 3D model of the fly ash, RFA particle, where *GS*, *ML* and *Q* designate glass, Mullite and Quartz, respectively

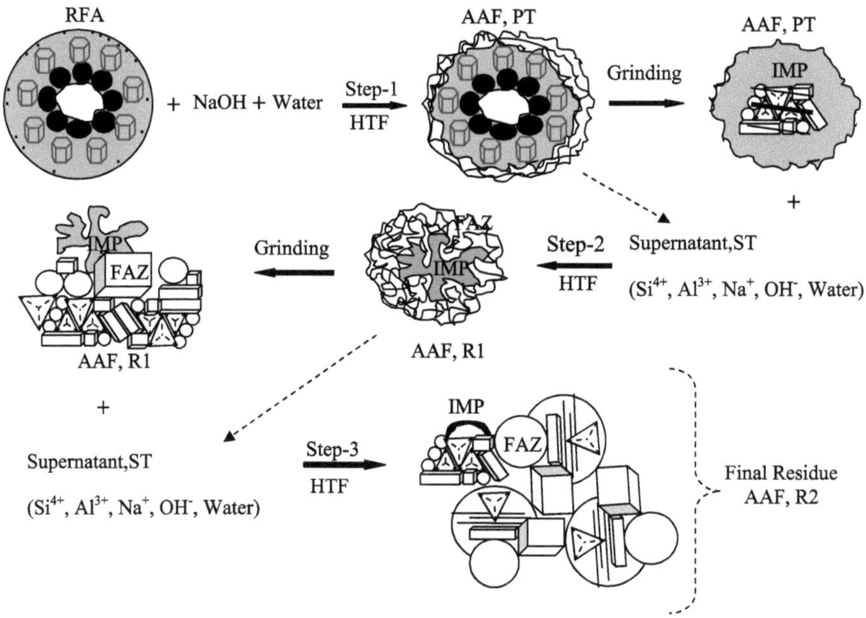

Fig. 6.18 A conceptual model of the process to explain step wise purification of the fly ash zeolites (*FAZ*) by hydrothermal treatment of fly ash (*HTF*), where *black dots*, *hexagons*, *deep black ovals*, *white inner most portion* and *white regular shapes*, IMP represent glass, Quartz, Mullite, hollow core of the RFA, FAZ and activated remains of the fly ash particles as impurities, respectively

polycrystalline shapes (viz., balls in 75–100 nm diameter of Hydroxy-sodalite, thickened rods and honeycombs of Na-P1) of FAZ (refer Fig. 6.14). Thus, in agreement to Wilkin and Barnes [30], Fig. 6.18 demonstrates that polycrystalline fly ash zeolites get crystallized in a supernatant due to solubility of unstable zeolites in alkaline environment, which have low SiO_2/Al_2O_3 ratio, after Step-1 (refer Table 6.7). A significant increase in this ratio, due to Step-2, confirms increased stability and high *CEC* of the residue, as well (refer Fig. 6.12). From Fig. 6.15 the residue of Step-3 can be assumed to be enriched in the more thermally stable form of zeolites FAZ at temperature lower than 600 °C. Mineral phase transition beyond this temperature in the TG-DTA curves can be attributed to conversion of Quartz and Mullite impurities to metastable zeolites.

With this view, grinding of the product residue of each step of the treatment, as described in the Fig. 6.18, can contribute significantly in exposing the surface of the impurities in the AAF. Moreover, the supernatant (i.e. once or twice used NaOH solution) of each step, provides free Si^{4+} and Al^{3+} ions in dissolved state, which can interact and precipitate as zeolitic compounds. Based on this hypothesis and the observation of the morphologies in Fig. 6.14a–f, it can be demonstrated that by recycling the AAF and its corresponding **S**, the inactivated portion of the fly ash residue (e.g., Quartz and Mullite in the core region of the RFA), can be dissolved by their alkali activation and by resorting to recycles R1 and R2.

6.3 Quantification of Transitions in Fly Ash-Zeolite and NaOH-Water Systems

Significant interaction between the fly ash and aqueous NaOH is now an established fact. Accordingly, hydrothermally treated mixture of the fly ash-NaOH-H_2O would yield two different systems: (a) matrix of the fly ash-zeolite-NaOH-H_2O system (the residues) and (b) NaOH-Si^{4+}-Al^{3+}-H_2O system (the supernatant). However, not many efforts have been made to quantify all possible transitions occurring in these systems when interaction occurs in a residue-supernatant (the filtrate) system [19]. Such transition could be (i) loss in Na^+, *pH* and *EC*, in the supernatant (ii) gain in Si^{4+} and/or Al^{3+}, in the supernatant, (iii) loss in SiO_2, Al_2O_3 in the residues and (iv) gain in Na_2O and zeolites, in the residues. With this in view, a study was initiated to monitor these transitions in the RFA and the residues AAF. It has been observed that Step-3 results in the highest transition in the residues, corresponding to 0.5 and 3 M NaOH for treatment duration varying from 24 to 36 h. Also, based on the FEG-SEM analyses (refer Fig. 5.10a–k), formation of newly synthesized particles, which vary in size from 30 to 2346 nm, has been confirmed. Such a wide range of particle sizes conforms to polycrystalline zeolites (viz., Na-P1, Faujasite, Chabazite, Hydroxy-sodalite and Analcime) [10], which have significant growth in ratio of oxygen to total of Si and Al atoms. This also confirms remarkable enlargement of zeolite rings from D-4R (i.e., four membered ring structure of rod or prismatic shaped crystal) to D-8R (octahedral shaped crystals).

A correlation of transitions of *pH* (refer Fig. 6.19) and electrical conductivity, *EC* (refer Fig. 6.20) elucidates some interesting effects of the three-step activation process. These figures exhibit that a reduction in both *pH* and *EC* of the supernatant influences some critical residues (viz., 0.5-PT-12, 1.5-PT-12, 3.0-PT-12 and 2.0-PT-36) of Step-1, 0.5-R1-12, 0.5-R2-12 and 1.5-R1-12 of Step-2, 1.5-R2-12,

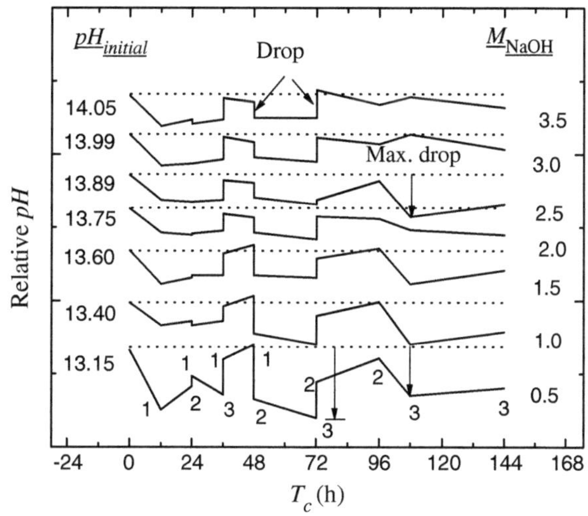

Fig. 6.19 Variation of *pH* of the supernatant with cumulative durations, where 3 against 36 h cumulative duration, represents three steps of treatment, each of 12 h duration

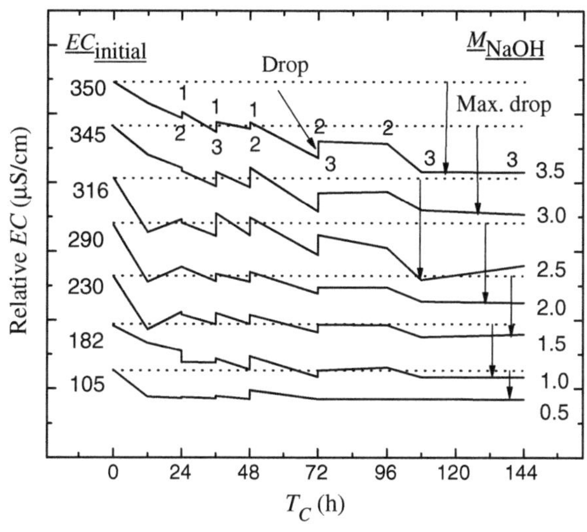

Fig. 6.20 Variation of *EC* of supernatant with cumulative durations, where 2 against 24 h cumulative duration represents two steps of treatment, each of 12 h duration

2.5-R2-24, 1.5-R2-36, 3.5-R2-36 and 3.0-R2-48 of Step-3. This can be attributed to variation of α_{Si} and α_{Al} (i.e., percentage relative change of Si and Al concentration in the supernatant, respectively) and variation of γ_{Si} and γ_{Al} (i.e., percentage relative change of SiO_2 and Al_2O_3 content in the residues, respectively), as depicted in Fig. 6.21. Further, the decrease in *pH* and *EC* can also be attributed to loss in alkalinity and migration of Na^+ and other cations, respectively, from the filtrates (supernatant, ST) to the residues.

Based on these variations in the ST and the residues, the X-ray diffraction patterns and FEG-SEM micrographs of the critical residues are presented in Figs. 6.22 and 6.23, respectively.

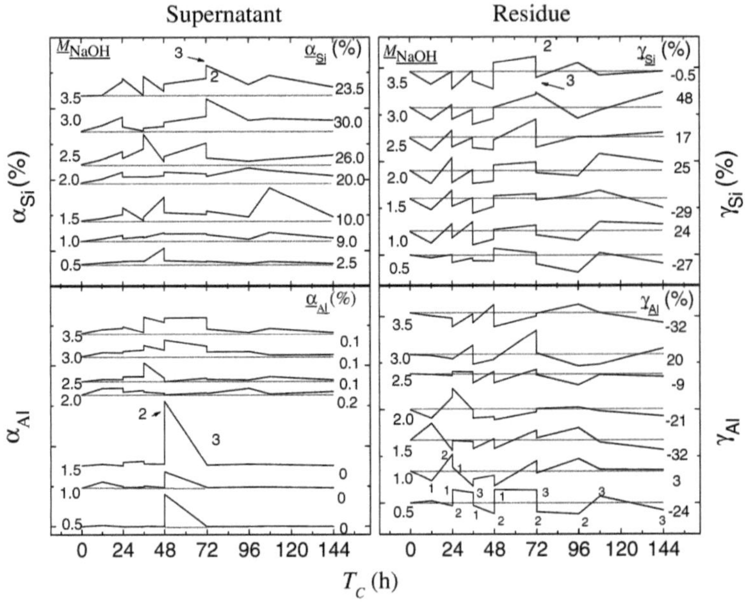

Fig. 6.21 Variations in Si and Al in the supernatant and residue, with cumulative duration of the treatment, where 2 against 48 h cumulative duration represents two steps of treatments, each of 24 h duration

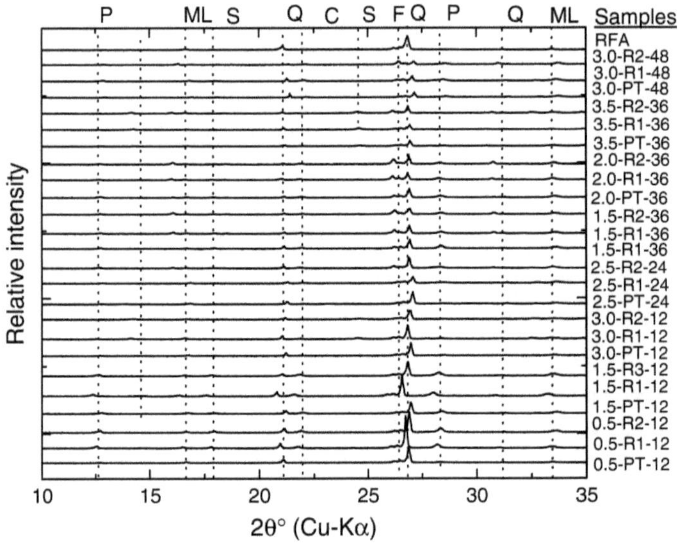

Fig. 6.22 XRD patterns of the fly ash and the residues

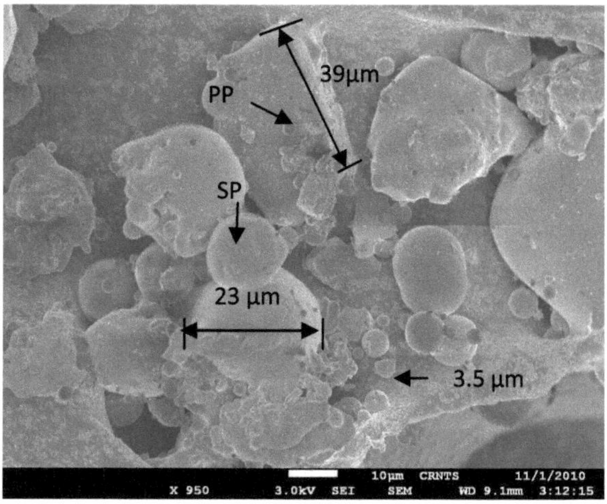

Fig. 6.23 A FEG-SEM micrograph of the hopper ash (RFA), where *PP* and *SP* designate *pillar shaped* and *spherical shaped* particles

It can be noticed from Fig. 6.22 that wide alteration in the mineralogy of the residues takes place and this results in the dissolution of peaks (viz., Quartz and Mullite, designated as Q and ML, respectively), which become boon for the formation of zeolitic minerals (viz., Na-P1, Chabazite, Hydroxy-sodalite and Faujasite designated by P, C, S and F, respectively).

Such developments in the fly ash (refer Fig. 6.23) commensurate with deposition of zeolites (rod, fibers, star and ball shaped Na-P1 in 0.5-PT-12, 1.5-PT-12, 3.0-PT-12 and 2.0-PT-36, respectively), as noticed in Fig. 6.24.

Further from Fig. 6.24 it can be observed that there is a wide variation in crystal sizes (e.g. from 30 to 2346 nm) of the residues of the TSA. To compare the FAZ with corresponding grade of pure zeolites, the characteristics of such zeolites are presented in Table 6.9. It can be noted from Fig. 6.24 that FAZ gets morphologically transformed during recycling treatment, from spherical (designated as *Z*13, Z14 in Table 6.10, which comprises of six membered ring crystal structure, D-6R) to cubic (designated as *Z*19-22 in Table 7.10, which comprises of four membered ring crystal structure, i.e. D-4R) and rhombohedral (designated as Z16 in Table 6. 10, and with crystal structure D-4R), and later to bi-pyramidal shapes (designated as Z24 in Table 6.10, which comprises of eight member ring crystal structure i.e., D-8R). Such transformation elucidates significant enlargement in zeolites' framework [O-Si-O-Al-O] of atoms, internal cage size and pores in the residues (refer Fig. 6.24), which in line with Thuadaij et al. [31].

Based on reanalysis of data from Table 6.10, Fig. 6.25a exhibits that Si/Al of most of the synthesized zeolites varies between 1 and 2.5, which corresponds to zeolites Na-P1, Analcime, Faujasite and Chabazite. All of these zeolites are having their O/T ratio approaching 2, with varying water molecules (refer Table 6.9). In fact, the Al/CT equal to 1 can be identified as a critical value when the surface negativity of the zeolitic minerals is thought to be neutralized by externally linked

Table 6.9 Characteristics of some pure zeolites from JCPDS files [1]

Zeolites	Designation	JCPDS code	Si/Al	Al/Na	O/T	$(H_2O)_n$	Shape	d_{100} (Å)
Analcime	A	19-1180	2	1	2	1	Orthorhombic	3.43
Chabazite	C	83-1498	2	1	2	–	Monoclinic	8.85
Clinoptilolite	CL	39-1383	5	2	2	24	Monoclinic	8.95
Erionite	E	39-1379	3	0.5	2	17	Hexagonal	2.85
Faujasite	F	12-0228	1.65	1	2	7	Cubic	5.71
Herschelite	H	19-1178	2	1	2	3	Rhombohedral	2.93
Hydroxy-sodalite	S	11-0401	1	0.75	2	1[a]	Cubic	3.63
Hydroxycancrinite	HC	28-1036	1.08	0.85	2.04	6	Hexagonal	3.26
Na-A	NA	43-0142	1.1	1	2	–	Cubic	12.29
Na-P1	P	39-0219	1.66	1	2	12	Monoclinic	3.17
Mordenite	MOR	06-0239	5	1	2	7	Orthorhombic	3.22
Mullite	ML	15-0776	0.33	0	1.6	–	Orthorhombic	3.39
Quartz	Q	83-2466	–	–	2	–	Hexagonal	3.33

[a]Number of OH molecule

Note A dash (–) represents not available, *T* is the sum of Si and Al and, *O* is the oxygen atoms present on the crystal surfaces, d_{100} is the d-spacing of the crystal lattice corresponding to 100 % relative intensity in X-ray diffractogram

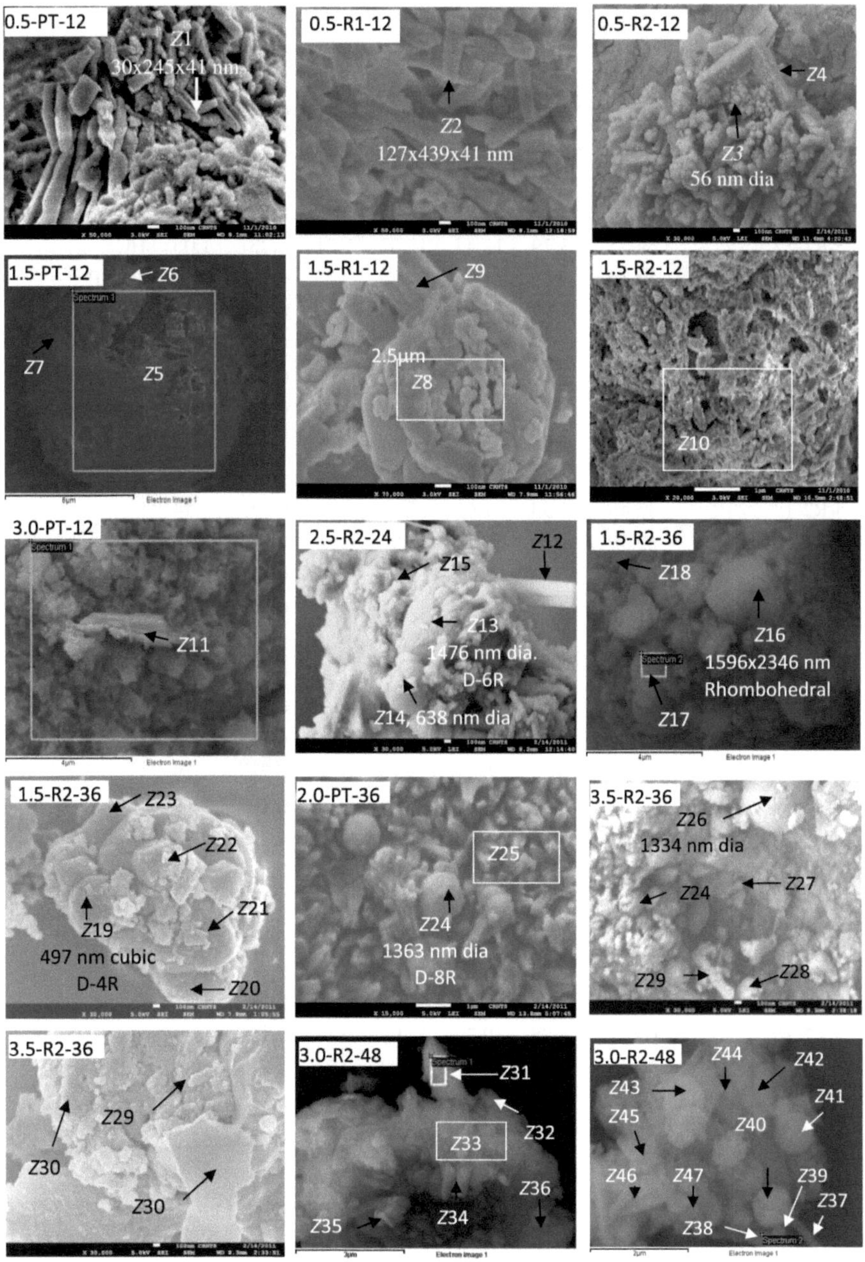

Fig. 6.24 FEG-SEM micrographs of the residues

cations (viz., Na^+, K^+, Ca^{2+}) [32–35]. Accordingly, a value of Si/Al ratio equal to 1, represents 50 % replacement of Si^{4+} by Al^{3+} in these zeolites, which is an indication of a superior zeolite [36, 37].

Table 6.10 Chemical transformation in the samples after alkali activation

Sample	Crystal	Elements detected in EDS analysis (atomic %)						Si/Al	Zeolite/mineral
		O	Na	Al	Si	K	Ca		
RFA	SP	67.65	–	11.61	17.74	0.27	0.83	1.53	Q
	FP	60.42	–	11.81	22.15	0.31	1.19	1.88	ML
0.5-PT-12	Z1	78.73	4.04	7.92	8.46	0.20	0.29	1.07	HC
0.5-R1-12	Z2	69.25	3.79	12.43	13.07	0.21	0.23	1.05	HC
0.5-R2-12	Z3	69.19	4.06	11.16	14.46	0.20	0.33	1.30	NA, P
	Z4	67.22	5.77	7.30	18.33	0.37	0.18	2.51	F
1.5-PT-12	Z5	80.76	3.29	9.55	5.57	–	0.27	0.58	Gel
	Z6	68.17	2.45	5.32	3.27	–	0.2	0.61	S, Gel
	Z7	54.98	1.63	10.85	4.97	–	0.17	0.46	Gel
1.5-R1-12	Z8	41.19	2.66	6.4	9.8	–	1.3	1.53	P
	Z9	54.07	3.25	14.84	14.82	–	1.55	1.00	HC
1.5-R2-12	Z10	45.45	1.38	1.65	5.67	0.11	0.65	3.44	E
3.0-PT-12	Z11	62.19	9.79	9.11	16.94	0.14	0.33	1.86	P
2.5-R2-24	Z12	67.96	15.93	6.57	8.14	–	0.44	1.24	S
	Z13	72.29	7.32	6.70	12.87	–	0.30	1.92	P
	Z14	67.90	9.27	8.04	13.35	–	0.41	1.66	S
	Z15	72.34	6.40	6.69	12.50	0.09	0.37	1.87	P
1.5-R2-36	Z16	67.27	8.18	6.77	16.46	–	0.13	2.43	F
	Z17	65.25	7.88	7.28	18.56	–	0.14	2.55	F
	Z18	63.70	6.80	8.46	19.39	0.22	0.28	2.29	C
	Z19	69.45	5.68	7.29	16.64	–	0.19	2.28	F
	Z20	65.69	5.34	7.35	20.03	–	–	2.73	F
	Z21	69.02	6.95	8.19	14.97	–	0.41	1.83	P
	Z22	70.82	6.66	6.45	14.85	–	0.11	2.30	F
	Z23	73.69	4.61	6.20	14.40	–	–	2.32	F
2.0-PT-36	Z24	65.39	9.22	8.62	16.02	–	–	1.86	P
	Z25	67.76	8.82	8.60	13.56	–	0.15	1.58	P
3.5-R2-36	Z26	65.49	5.85	10.79	14.99	0.17	0.15	1.39	F
	Z27	57.94	5.81	16.27	17.03	0.21	0.21	1.05	S
	Z28	63.15	6.45	12.90	14.42	0.23	0.21	1.12	S
	Z29	61.42	7.20	12.89	13.92	0.25	0.35	1.08	S
	Z30	62.42	14.15	10.71	11.78	–	0.23	1.10	S
3.0-R2-48	Z31	74.20	3.56	0.98	21.04	–	–	21.47	β or Q
	Z32	72.33	12.11	2.28	13.29	–	–	5.83	MOR
	Z33	67.26	14.34	5.55	12.10	0.16	0.20	2.18	F
	Z34	66.35	11.85	7.25	13.82	0.08	–	1.91	P
	Z35	62.64	9.59	8.81	18.08	–	–	2.05	A
	Z36	57.84	12.50	9.13	19.02	–	–	2.08	A

(continued)

Table 6.10 (continued)

Sample	Crystal	Elements detected in EDS analysis (atomic %)						Si/Al	Zeolite/mineral
		O	Na	Al	Si	K	Ca		
	Z37	67.22	23.62	2.67	6.49	–	–	2.43	F
	Z38	27.22	3.34	0.86	1.60	–	–	1.86	P
	Z39	66.38	14.56	6.24	12.32	–	–	1.97	P
	Z40	67.63	10.79	7.60	13.60	0.13	–	1.79	P
	Z41	69.23	9.42	6.96	12.28	–	0.14	1.76	P
	Z42	66.86	15.30	5.93	10.58	–	0.13	1.78	P
	Z43	68.38	13.93	4.23	12.48	–	0.12	2.95	F
	Z44	65.13	18.39	5.42	10.18	–	0.11	1.88	P
	Z45	65.44	16.24	5.86	10.96	–	0.21	1.87	P
	Z46	61.01	20.67	6.16	11.42	–	0.17	1.85	P
	Z47	62.58	15.03	7.40	14.15	–	0.16	1.91	P

– Could not be detected
Note Refer to Table 7.9 for details of designation for zeolites, listed in the last column of this table

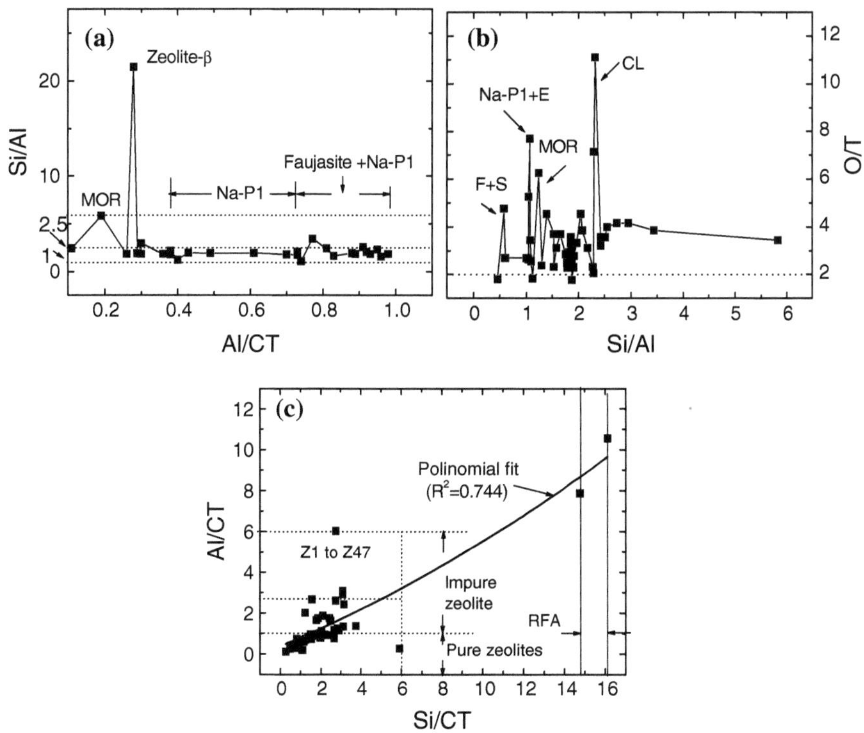

Fig. 6.25 Correlations between characteristic ratios **a** Al/CT versus Si/Al, **b** Si/Al versus O/T and **c** Si/CT versus Al/CT, where CT is sum of cations (viz., Na^+, K^+, Ca^{2+}), T is the sum of Si and Al and, O is the oxygen atoms present on the crystal surfaces, observed by the energy dispersive analysis, EDS of the micrographs

On the other hand, a wide variation in Si/Al from 1 to 21, as noticed in Fig. 6.25a, reveals impure zeolites, which have less Al^{3+} and significantly increased Si^{4+}, which can be in the forms of Quartz and Mullite, or a high silica zeolite β (not detected in the XRD) as exhibited in Fig. 6.22. On the contrary, such a combination of ions may commensurate with an impure form of zeolites, as well. Further from Fig. 6.25b it is apparent that O/T (i.e., oxygen atoms/sum of Si and Al atoms) increases without any change in Si/Al ratio, at 1 and 2.5. This elucidates a situation when due to activation in three steps, more oxygen atoms are incorporated in the zeolites Na-P1 to Faujasite (refer Table 6.10) and a polymerization of [Si-O-Al] linkage occurs, which also indicate formation of more porous and pure zeolites, by undergoing several transitions in both the residue and the filtrate.

In addition, from Table 6.9 it can be observed that majority of pure zeolite crystals are orthorhombic, rhombohedral, and spherical in shapes, which are observed to have resemblance with the shapes noticed in the micrographs of the zeolitized residues (refer Fig. 6.24). Further, with reference to their less values of d-spacing of the crystal lattice (d_{100} = 2.85–3.26 Å, refer Table 6.9), these zeolites are inferred to have more closely packed atomic planes, which indicates their better structural stability, as compared to monoclinic and cubic shapes, formed in the residues of Step-1 treatment.

Accordingly, a general relationship developed between Si/CT and Al/CT, is noticeable, as depicted in Fig. 6.25c. This figure reveals that Al/CT increases with an increase in Si/CT. However, from Fig. 7.25c most of the zeolites (designated in Tables 6.9 and 6.10) are observed to have Si/CT and Al/CT ranging from 0–4 and 0–2.5, respectively, which are significantly below those for the fly ash, RFA. This shows high effectiveness of the adopted three-step activation process, a remarkable improvement over the previous researchers.

6.4 Formation of Meso- and Micro-pores by TSA

Researchers [36, 38] have characterized NaOH activated fly ash residues (zeolites) and have ascertained their highly zeolitic nature [7]. It has also been opined [37, 39] that the purity of these residues is mainly decided by their cation exchange capacity, mineralogy and morphology. However, study of pore characteristics (viz., size and volume) of such residues has not been explored in detail.

In this context, efforts were made to demonstrate the formation of pores and their effect on specific gravity, specific surface area and cation exchange capacity of three types of residues (produced by using NaOH aqueous solution and its spent solution (the filtrate), respectively). Based on N_2 gas absorption, isotherms and infrared transmittance spectra, residues obtained from second and third steps (each of 24 h) of the treatment by filtrates of 1.5 M NaOH were established to be significantly enriched in finer meso- and micro-pores, respectively, as compared to a pure and macro-porous zeolite 4A (zeolite RZP).

In agreement to the findings reported in the literature [33] it has been observed that there is an increase in specific gravity, *G*, and specific surface area, *SSA*, of the residues 1.5-PT-24 as compared to its counterparts (refer Table 6.11).

It can be noticed from the data presented in the table that Step-2 (the R1) results in supernatant activated residues, which have reduced *G* and increased *SSA* values, as compared to the residues of Step-1. These changes can be attributed to significantly increased pore volume from 0.01725 to 0.036 cm^3/g (increase by 108 %), as observed in Table 6.12. Incidentally, Step-2 treatment is also effective in (i) creation of smaller meso-pores (3–10 nm) by 6.94 % (ii) reduction in bigger meso-pores viz. 10–50 nm by 5.56 % and (iii) reduction in macro-pores by 1.4 %. Accordingly, such improvement in the pore structure obtained from Step-2 is responsible for remarkably enhancing cation exchange capacity, *CEC*, (refer Table 6.11). This can also be attributed to increase in *SSA* values, in general. However, *SSA* values obtained from the BET (N_2 gas relative pressure in between 0.05 and 0.3) correspond to external pores, while that computed by EGME adsorption method correspond to total of external and internal pores in the residues. Accordingly, Step-2 can be held responsible for creation of more internal pores.

Further from Table 6.13, quantities of SiO_2 and Al_2O_3 are found to increase due to Step-2 (the R1) treatment. In agreement to Shanten et al. [40], such chemical changes in the residues reveal that silica dissolved during Step-1 (the PT) in the filtrate is regained back during Step-2 treatment and in turn, there could be (i) an improved coordination between Si and Al tetrahedra and hence (ii) better pore size distribution in the residues, as apparent from Table 6.12. Furthermore, it is

Table 6.11 Comparison of various parameters of the fly ash (RFA), fly ash residues and reference zeolite RZP

Parameter	RFA	1.5-PT-24	1.5-R1-24	1.5-R2-24	RZP
G	2.18	2.94	2.63	2.55	2.14
SSA_{BET} (m^2/g)	2.83	7.04	7.38	6.23	1.87
SSA_{EGME} (m^2/g)	7.0	33.9	110.5	155.6	94.5
CEC (meq./100 g)	8	296	470	843	450

Table 6.12 Pore size distribution for fly ash, activated fly ashes based on BET analysis

Sample	Total pore volume (cm^3/g)	% Volume contribution of various pore sizes (nm)				
		<2	2–3.0	3.0–10	10–50	>50
		Micro	Meso			Macro
RFA	0.00195	0	0	0	35.9	64.1
1.5-PT-24	0.01725	–	–	–	33.33	66.67
1.5-R1-24	0.036	–	–	6.94	27.77	65.27
1.5-R2-24	0.0235	2.1	4.25	2.12	25.53	66.00
RZP	0.00325	–	–	–	–	100

– Below detection limit

Table 6.13 Chemical composition of various samples

Sample	Oxides (by wt%)											
	Al_2O_3	BaO	CaO	Fe_2O_3	K_2O	MgO	MnO_2	Na_2O	P_2O_5	SiO_2	SrO	TiO_2
RFA	26.05	0.14	1.88	5.14	0.66	0.39	0.16	0.054	10.5	63.85	0.25	1.52
1.5-PT-24	25.57	0.10	1.63	5.51	0.38	0.34	0.05	10.60	NA	48.71	0.04	1.51
1.5-R1-24	26.47	0.09	1.85	5.53	0.27	0.41	0.05	9.77	0.04	53.95	0.04	1.51
1.5-R2-24	34.55	0.09	1.86	3.60	0.30	0.39	0.05	8.57	0.05	49.13	0.04	1.40
RZP	46.04	0.03	0.22	0	0	0	0.02	16.87	0.02	36.57	0.01	0.24

interesting to notice that characteristic changes in the residues obtained from Step-3 (the R2), which results in an exceptionally high SSA_{EGME} value (refer Table 6.11), are indicative of formation of finer meso-pores (2–3 nm by 4.25 % by volume) formation. Most encouraging observation is that this step also creates some micro-pore volumes (2.1 %, refer Table 6.12), which can be attributed to very high *CEC* (up to 843 meq./100 g, which is higher than that of the reference zeolite RZP). In fact, the reason behind relatively low *CEC* of the RZP could be its less surface area *SSA* because of absence of micro/meso-pores and presence of macro-pores as highlighted in Table 6.12. Furthermore, from Fig. 6.26 it can be noticed that zeolite 4A (the RZP) exhibits more gas condensation (i.e., interpreted from the gap between adsorption and desorption isotherms) at higher pressure than the fly ash.

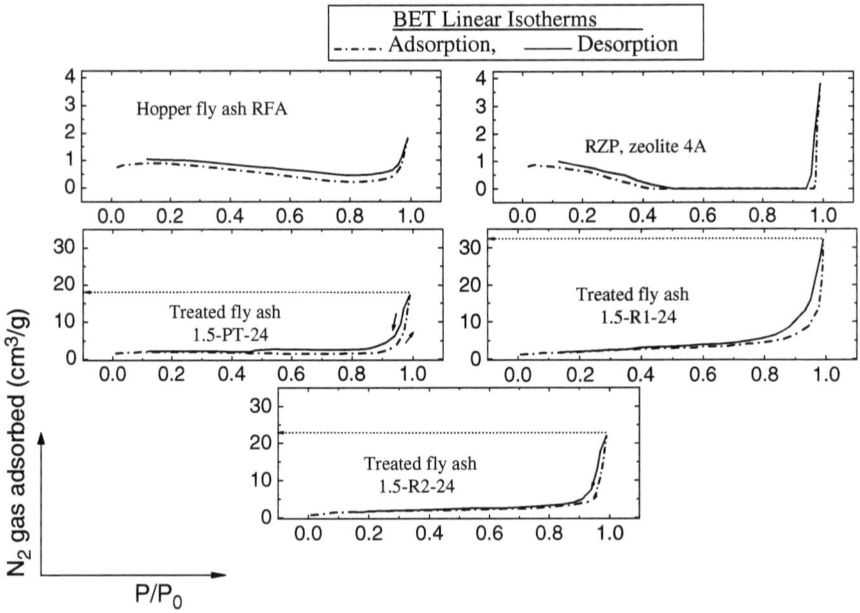

Fig. 6.26 Variation in N_2 gas absorption with relative gas pressure, P/P_0 maintained in the BET analysis

This can be indicative of less interconnected micro-pores in the residues (the treated fly ash). Also, presence of type III hysteresis [41–43] in sorption isotherms for residues 1.5-R1-24 indicates gas condensation in the capillaries between meso-pores. In addition, hysteresis lag between two curves (absorption and desorption) can be attributed to pore blocking (termed as ink bottle effect) during desorption in the residues. Such effects are relatively less noticeable in isotherms of residues 1.5-R2-24, which might be a revelation of their well interconnected micro- to meso-pores. More precisely, dominant pore diameter in these samples has been shown as black and red arrows on the dV/[dlog(D)] verses pore diameter, D, curves corresponding to both adsorption and desorption, respectively, (refer Fig. 6.27). From the figure it can be noticed that desorption curves terminate at lower pore diameters as compared to the adsorption curves. This can be attributed to (i) gas condensation on the pore boundaries, (ii) bigger pores i.e., meso-pores getting blocked, which results in a reduced dominant pore diameter from desorption trends. Such pore blocking effect is found to be quite intensive in case of zeolite 4A, in which desorption curve shifts remarkably to the smaller pores. Exceptionally, micro-pores are observed in residues 1.5-R2-24 only, in which negligible pore blocking occurs for D < 100 Å.

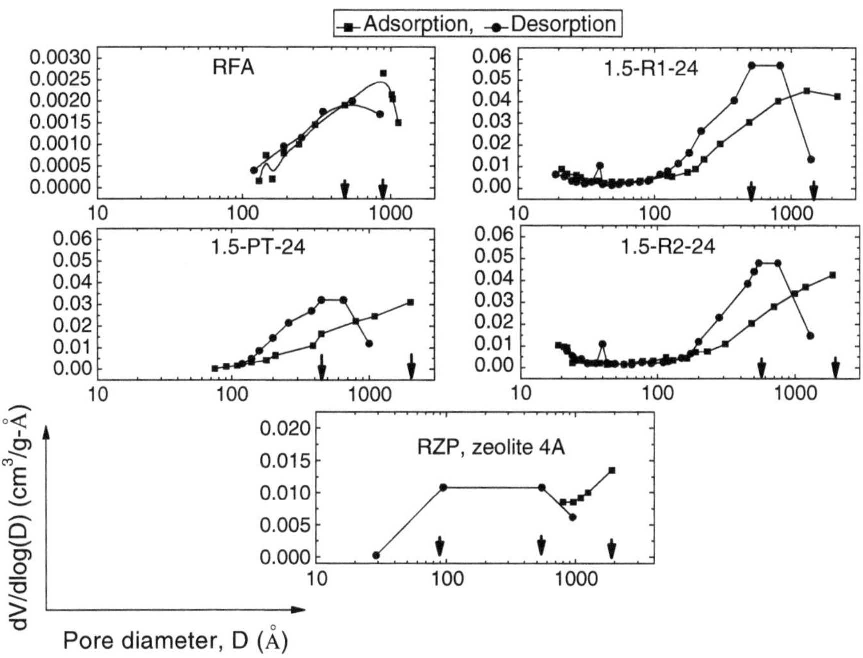

Fig. 6.27 Barrett-Joyner-Halenda (BJH) sorption curves for the fly ash, activated fly ash residues and reference zeolite 4A (*RZP*), where *black arrows* represent dominant pore diameters corresponding to N_2 gas adsorption and desorption

Fig. 6.28 FT-IR spectra of the fly ash RFA, its residues 1.5-PT-24, 1.5-R1-24, 1.5-R2-24 and RZP (zeolite 4A)

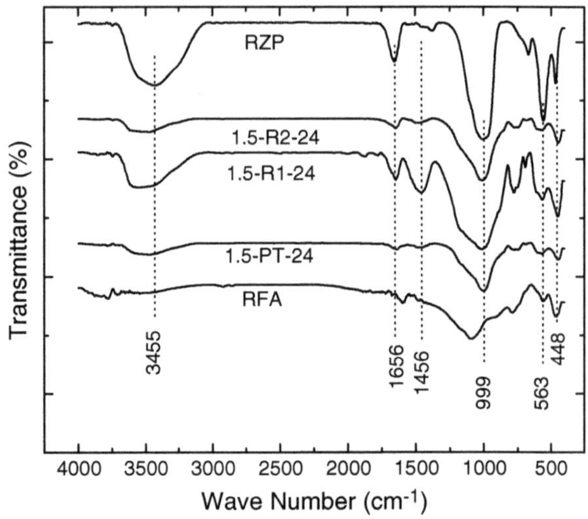

Useful information regarding pore size of these residues is also obtained from FT-IR spectra, as depicted in Fig. 6.28. It can be observed that the spectrum of zeolite 4A (designated as RZP) exhibits sharper bands than the residues 1.5-R1-24, which has infrared transmittance value up to nearly half of its counterpart at wave numbers viz. 3455, 1656, 999 cm^{-1}. Based on very sharp bands at 563 cm^{-1} it can be opined that RZP, zeolite 4A comprises of relatively more polymerized double ring framework than the residues 1.5-R1-24. On the contrary, pore openings represented by bands at 448 cm^{-1} [44] are similar for the residues 1.5-R1-24 and zeolite RZP. Surprisingly, zeolite RZP and residues 1.5-R2-24 display negligible band at 1456 cm^{-1} (i.e., indicative of carbon or nitrogen to hydrogen bonds), which might be beneficial for the residues to exhibit high *CEC* (refer Table 5.13). In addition, sharper band in the spectrum of the residue 1.5-R2-24 at 448 cm^{-1} (i.e., an indication of micro-pores), is found to be in agreement with pore volume, shown in BET results, as apparent from Table 6.12.

Figure 6.29 depicts morphological transitions in these samples. It can be noted that as compared to fly ash RFA (which exhibits matrix of bigger particles of Quartz, Mullite, small spherules of glass and cenospheres), residues 1.5-R2-24 (which are agglomerates of spherules and cubic to cuboidal shaped crystals, conforming to those in micrographs of zeolite RZP) exhibit significant changes in the particle morphology and pores (the opening between crystals and within their framework also). Similarity between the zeolite RZP and residues 1.5-R2-24 can also be established from their XRD diffractograms (refer Fig. 6.30), which exhibit many mineral peaks conforming to zeolite 4A. In addition, minerals (viz., Quartz and Mullite) present in RFA are found to get converted to another zeolite Na-P1. The difference of 2θ° (between peak positions) elucidates atomic level transition in these samples and hence alteration in their pore characteristics, as well.

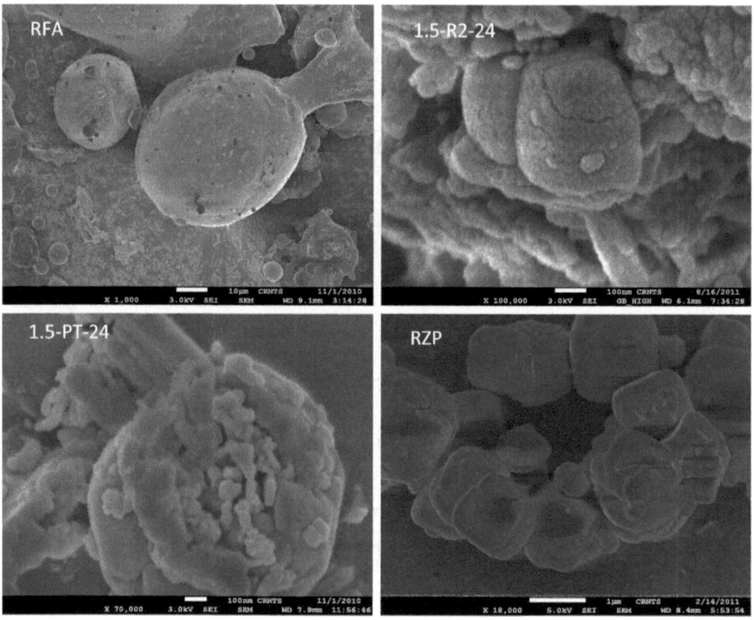

Fig. 6.29 FEG-SEM micrographs of the fly ash RFA, its conventional residue 1.5-PT-24, most porous residue 1.5-R2-24 and RZP (zeolite 4A)

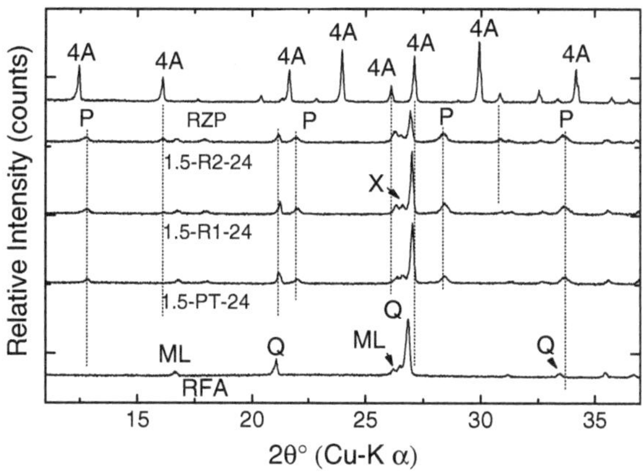

Fig. 6.30 XRD diffractogram of the fly ash RFA, its residues 1.5-PT-24, 1.5-R1-24, 1.5-R2-24 and reference zeolite (*RZP*), where *4A*, *P*, *X*, *ML* and *Q* designate zeolites 4A, Na-P1, Na-X, Mullite and Quartz, respectively

6.5 Thermal Stability of Superior Fly Ash Zeolites Synthesized by TSA

It can be noticed from TGA curve, depicted in Fig. 6.31a, reproduced here for analysis that the FAZs, 1.5-R2-24 and 1.5-R2-48 undergo maximum weight loss (up to 9.12 %), as compared to other AAFs and the fly ash, when heated up to 100 °C.

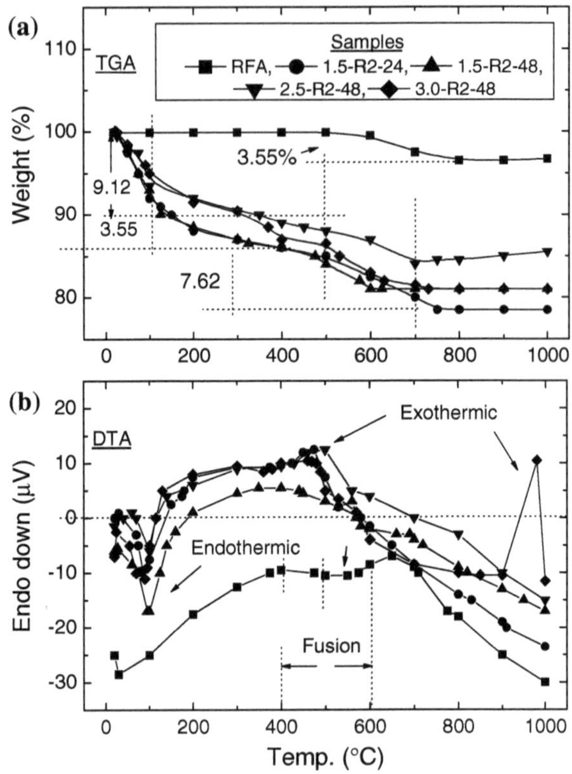

Fig. 6.31 Variation in response of the FAZ under thermal analyses **a** TGA and **b** DTA

This could be due to rapid and increased dehydration of monohydrate (1 H_2O) from increased surface pore in these zeolites, which can be confirmed by an endothermic peak in the DTA curve, depicted in Fig. 6.31b. This also supports the maximum *CEC* of 1.5-R2-24 and exhibits highly porous structure. Further heating (at 100–500 °C) of the sample causes more reduction (by 3.55 %) in the weight. This can be attributed to calcination of unburnt carbon in the ash and escape of all organic gases (CO and CO_2) from the sample. Also, at elevated temperature from 500 to 1000 °C, a loss in weight by 7.62 % is an indication of mineral phase transition and decomposition of oxides (Na_2O, CaO, MgO and Fe_2O_3).

6.6 Synthesis of High Grade Zeolite by TSA-Fusion

Several researchers have established fusion of fly ash-NaOH mixture, at 450–900 °C. Accordingly, the fly ash gets partially converted to zeolite X when the fused ash is further treated in a hydrothermal system. However, such a solid residue (the end product) exhibits low cation exchange capacity. Such deficiency in the zeolite can be attribute to improper contact between the NaOH and the fly ash particles, during their fusion. In such a situation, to enhance degree of fusion, the present study attempts to demonstrate a three-step fusion technique, which does not favor any further treatment in a hydrothermal system. Observed changes in the characteristics (viz., cation exchange capacity, pore volume, pore area, mineralogy, and thermal stability) at the end product of Step-3 fusion corresponding to NaOH/fly ash ratio equal to 1.0 proves suitability of this fusion technique for synthesis of better grade of zeolite-X viz. Na-X and Na-Y.

6.6.1 Need of Reactivation by Fusion

A smooth fly ash particle (refer Fig. 6.32a) has been frequently characterized and established that it can undergo fusion reactions and get covered by their products (refer Fig. 6.32b). Conceptually, Fig. 6.32c, d exhibits their pictorial models where glass, Quartz and Mullite are supposed to be located on the outer, intermediate and inner region of the fly ash particle, respectively [7, 10, 28]. As demonstrated by Fig. 6.32c, d, upon a sudden increase in temperature up to 500 °C in the Muffle furnace, a temperature gradient between outer and inner portion of the particle would get established and this will affect the degree of fusion. Depending upon wide variation in values of thermal conductivity of glass (1.5 W/m K), Quartz (3.5 W/m K) and Mullite (6 W/m K) [44], sufficient duration will be elapsed in equalizing the temperature from outer to inner region of the particle (refer

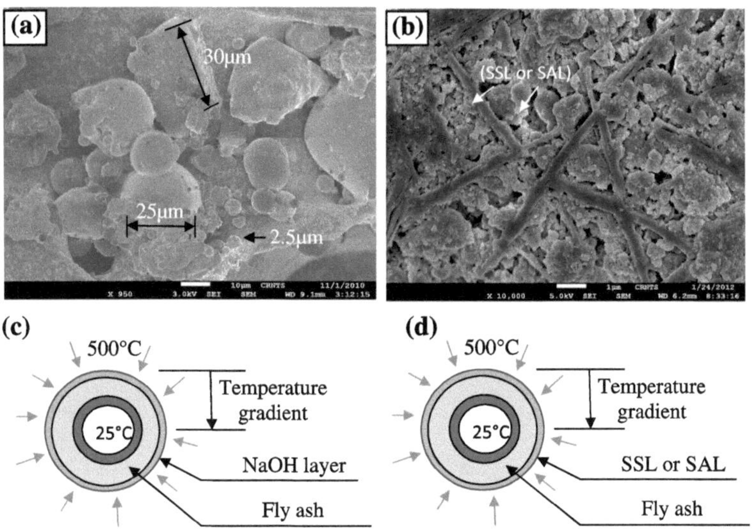

Fig. 6.32 Surface properties of **a** fly ash, **b** conventional fusion product obtained after Step-1 for NaOH/RFA ratio = 1.0 designated by 1.0-F1, **c** conceptual models of NaOH-fly ash system before fusion and **d** conceptual models of NaOH-fly ash system after fusion

Fig. 6.32c) [45]. In the early phase of this duration, NaOH would melt and interact with glass during fusion F1 [8, 28, 44, 45]. In fact, the scenario in the conventional fusion technique favors high resistance to flow of thermal flux through glass, and hence would elapse maximum duration in fusing it. This could be responsible for negligible activation of Quartz and Mullite [46]. In addition, deposition of fusion outputs viz. sodium silicate layer (SSL) and/or sodium aluminate layer (SAL) on the residual RFA particles (refer Fig. 6.32b) [20, 21].

As a result of hardening of these layers, further fusion of the residual fly ash with the molten NaOH would get restricted. This drawback of the conventional fusion process still remains unattended. In this situation, grinding of a fused residue manually is warranted to break the SSL and SAL and ascertain their mixing with the residual fly ash, before the second and third steps of the fusion. To investigate the extent of fusion of the RFA, the characterization (viz., physico-chemico-mineralogical, morphological) of all the residues was also essentially conducted.

6.6.2 *Evaluation of Major Transition in the Residues of Three-Step Fusion*

It is apparent from Fig. 6.33a that three-step fusion of fly ash did not yield considerable loss in weight of the sample and majority of such loss occurs in Step-1 fusion (the F1). This observation is in line with negligible variations in major oxides after Step-2 (the F2) and Step-3 (the F3) of fusion (refer Fig. 6.34).

However, a shift in various peaks present in the XRD diffractograms (refer Fig. 6.35) from lower to higher X-ray scattering angle (i.e., inversely proportional to d-spacing of a mineral) is an indication of a novel mineralogical phase transition, which has not been reported earlier from conventional fusion followed by hydrothermal method.

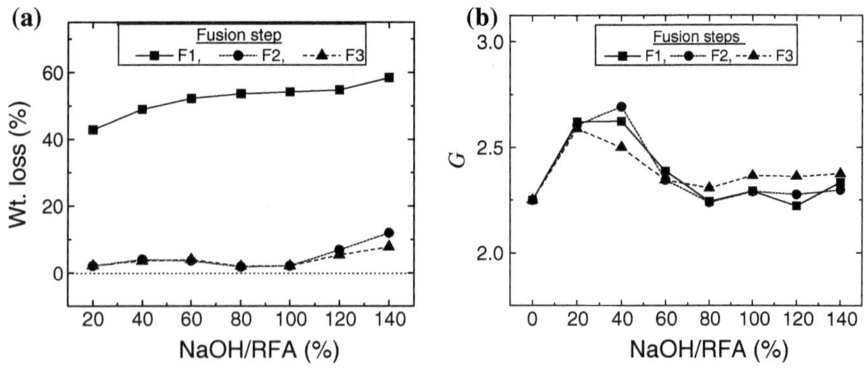

Fig. 6.33 Variation in the fusion residues **a** weight loss and **b** specific gravity *G*

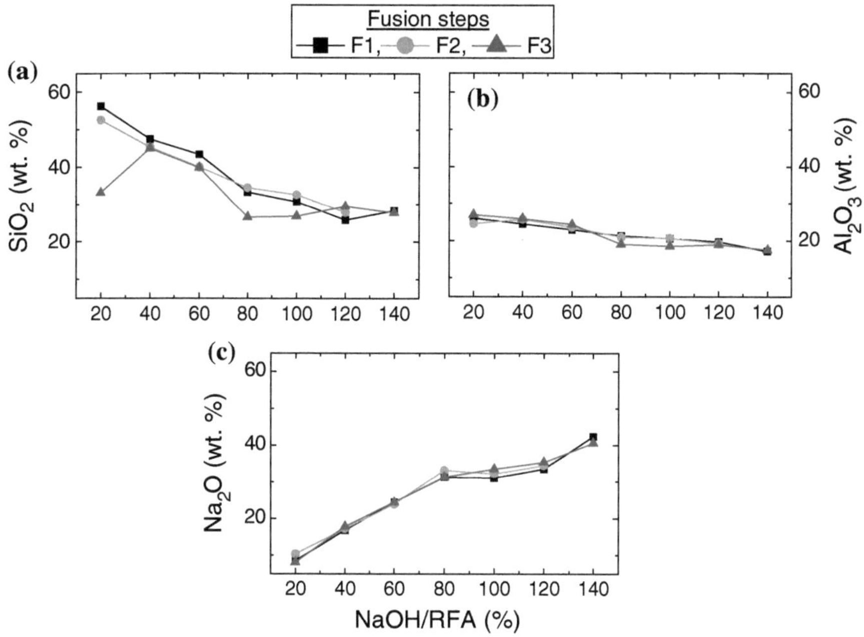

Fig. 6.34 Variation of major oxides **a** SiO_2, **b** Al_2O_3 and **c** Na_2O in the residues

In comparison to XRD diffractogram of the fly ash (refer Fig. 6.35a), such shift in the prominent peaks is noticeable from 2θ = 27.01 to 29.68, each of which conform to two different grades of zeolite X (viz., Na-X and Na-Y). This could be an indication of decrease in d-spacing and hence, increased atomic packing in the zeolite X due to reactivation during second and third steps of fusion, corresponding to NaOH/RFA ranging from 1.0 to 1.4.

In addition, irrespective of higher *CEC* of the fusion products corresponding to higher value of NaOH/RFA ratio (=1.4). Figure 6.36 hints similar improvements in some fusion products obtained by using lower values of NaOH/RFA ratio viz. 0.4 and 1.0, which results in an enhanced *CEC* of the fused fly ash due to fusion during Step-2 and 3. This can be attributed to (i) development of surface as well as structural negativity, (ii) creation of more pores and (iii) formation of zeolitic minerals. In this context it can be noticed from Fig. 6.37 and Table 6.14 that Si/Al ratios commensurate to zeolite Na-X corresponding to EDS spectrums-2, 3 and 4 and zeolite Na-Y corresponding to spectrums-1 and 5 [1], present in fused fly ash 1.0-F3.

Furthermore, from Table 7.15 and BET analysis results (refer Figs. 6.38 and 6.39), Step-3 fusion of the fly ash yields significant increase in the pore characteristics (viz., area, diameter and volume) of the resulting products corresponding to NaOH/RFA ratios viz. 0.4 and 1.0. It can be noticed that the pore volume increases drastically up to 0.0285 cm^3/g (i.e., 1361 % of that for fly ash) for 1.0-F3.

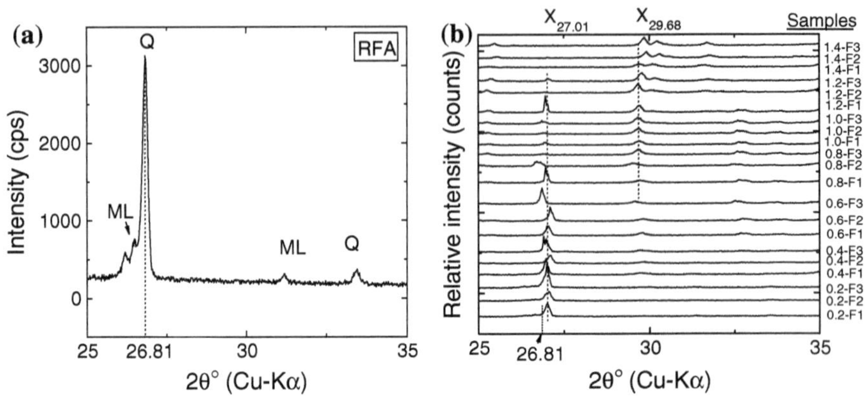

Fig. 6.35 XRD diffractograms of **a** fly ash RFA and **b** fused fly ash residues

Fig. 6.36 Variation in *CEC* of fusion products with NaOH/RFA ratios

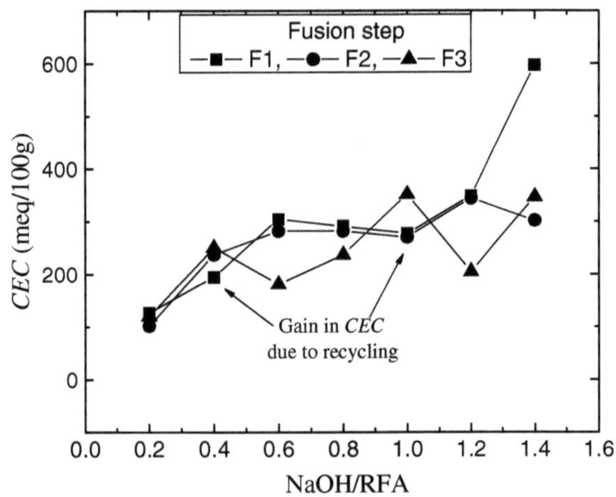

In fact, such increase in the pore volume should be responsible for a reduction in the specific gravity of the residues of Step-2 and 3 fusions. However, Fig. 6.33b reveals an increase in *G* with increasing fusion steps. This could be possible when volatile gaseous products get entrapped in meso- (i.e., 2–50 nm) to macro-pores (i.e., >50 nm) present in the residues 1.0-F3 [41].

Further, in agreement to increased *CEC*, this product also exhibits (i) remarkable increase in the meso-pores of size ranging from 2.5 to 10 nm and (ii) formation of many macro- pores (>50 nm, refer Table 6.15). Such results also confirm the formation of higher grades of more porous zeolitic products by the adopted fusion method. In addition, the dominant pore diameter in this product is noticed to be 30 nm followed by 6.5 nm, which are also noticed in its two micrographs (upper and lower images in Fig. 6.38 at resolutions of 300,000 and 30,000, respectively).

Fig. 6.37 EDS spectrums of the residue, 1.0-F3

Table 6.14 Elements present on the surface of the crystals within the EDS spectrum

Spectrum	Atomic (%)								Si/Al
	O	Na	Al	Si	Ca	Ti	Fe	Cu	
1	52.29	43.19	1.92	2.6	0	0	0	0	1.35
2	51.03	21.17	3.86	10.05	12.08	0.21	0.36	1.25	2.6
3	46.47	30.41	5.58	13.66	3.88	0	0	0	2.44
4	44.37	43.25	3.4	6.37		0	0	2.6	1.87
5	47.27	44.85	1.66	2.9	0.69	0	0	2.63	1.74

These pores are larger in fused ash 1.0-F3 than 0.4-F3, which has majority of pores of 2.5 nm size (refer Fig. 6.39).

The superiority of fusion product 1.0-F3 and its macro-porous zeolites (refer upper micrograph in Fig. 6.38) is again established from its TGA/DTA curves (refer Fig. 6.40, reproduced here for analysis). A large loss of weight (≈30 %) (refer Fig. 6.40a) and a sharp endothermic down spike (refer Fig. 6.40b) reveals its strongly hydrophilic characteristics and escape of all volatile gases responsible for increased *G* values (refer Step-2 and 3 curves in Fig. 6.33b), more surface and structural water holding capacity, which are responsible for its higher *CEC* (refer Fig. 6.36) and also validates its highly porous characteristics (refer Figs. 6.37 and 6.38). In addition, negligible weight loss in the TGA curve and absence of exothermic peak in the DTA curve of fused ash 1.0-F3 confirms its thermal stability up to 700 °C beyond which both the TGA and DTA curve of this product make it inferior to its counterpart 0.4-F3, the fused ash.

Table 6.15 Pore size distribution for fly ash, fused ash based on BET analysis

Sample	Total Pore volume (cm^3/g)	Pore volume (%) contribution of various sizes (nm)				
		<2	2–2.5	2.5–10	10–50	>50
RFA	0.00195	0	0	0	35.89	64.1
0.4-F3	0.016	0	3.75	4.38	31.25	37.5
1.0-F3	0.0285	0	0	12.2	30.7	68.42

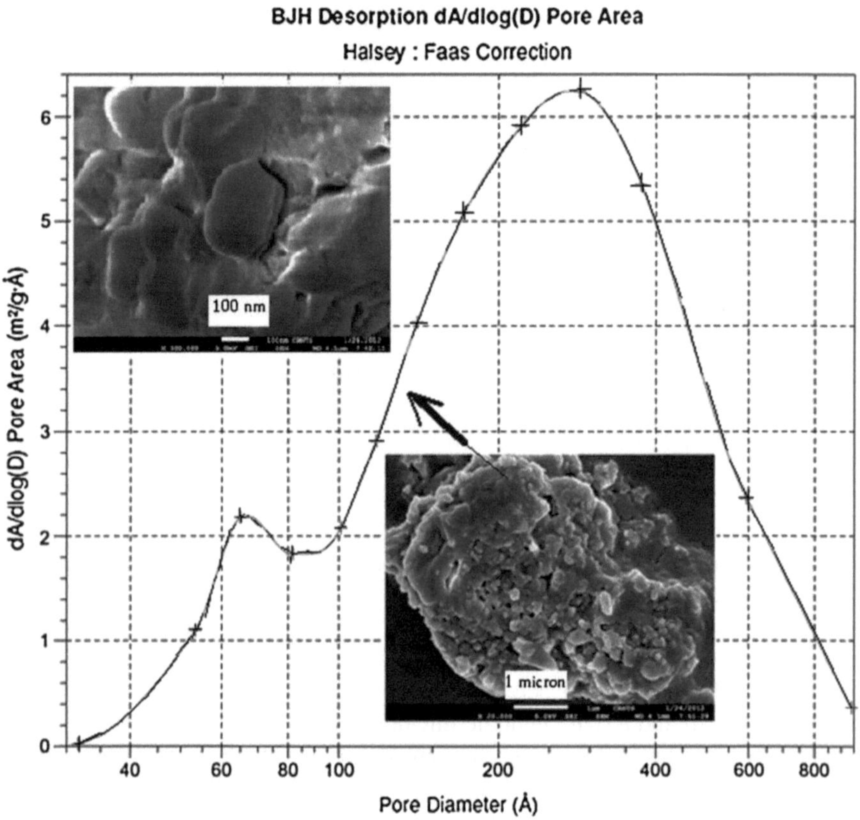

Fig. 6.38 Pore size distribution in fused fly ash 1.0-F3

6.7 Comparative Study of TSA Products Obtained from Hydrothermal and Fusion Methods

The products of both the TSA by resorting to hydrothermal and fusion techniques are significantly different as established above. Moreover, attempts have been made in the following section to establish superiority between the most superior products

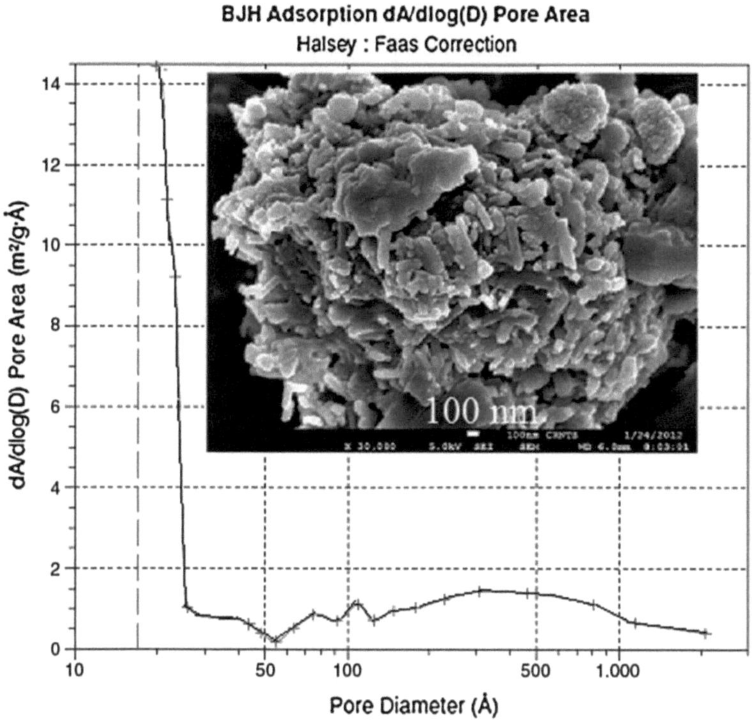

Fig. 6.39 Pore size distribution in fused fly ash 0.4-F3

obtained from both the techniques. In order to look in the micro-structural bond level in the products, FTIR analysis was carried out as described below.

6.7.1 *Fourier Transform-Infrared (FT-IR) Analysis*

FT-IR spectra of the RFA (the hopper ash), superior residues of the TSA (i.e., both hydrothermal and fusion method) and the commercial grade, reference zeolite 4A powder (RZP), are presented in Fig. 6.41. From this figure it is apparent that the residues of TSA are having more or less similarities with the RZP zeolite at many FT-IR bands in terms of their wave numbers (viz., 3466.3, 1650.3, 1008 and 541.3 cm^{-1}). On the contrary, the bands at 2974, 1473.14 and 859.8 cm^{-1} become decisive wave numbers of importance between the RZP and the residues of the TSA.

It can be noticed from Fig. 6.38 that the band at 2974 cm^{-1} corresponds to asymmetric stretching vibration of C-H bonds and this, as an impurity, is present in most of the residue except 1.5-R2-24. Another bands at 1473.14 cm^{-1} (i.e., indicative of carbonates and C-H bonds, as impurities) is absent from the standard

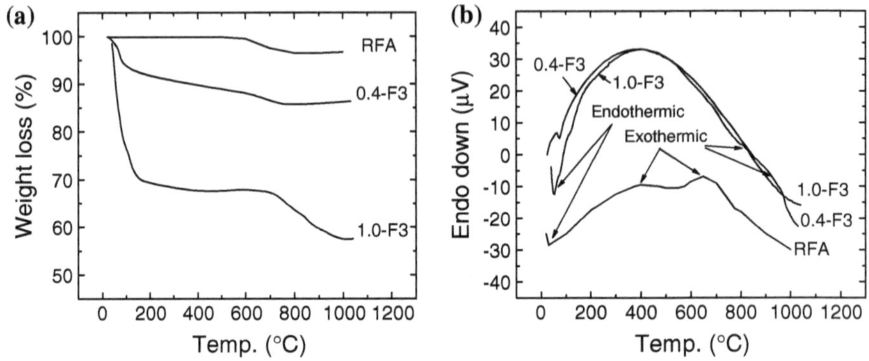

Fig. 6.40 Thermal response of the raw fly ash, fused fly ashes and the standard zeolite RZP by **a** TGA and **b** DTA analyses

Fig. 6.41 FT-IR spectra of superior products of the TSA

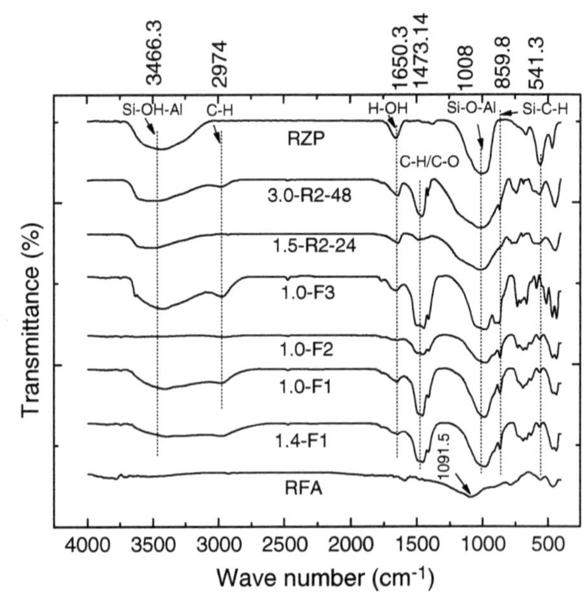

zeolite 4A (designated by RZP) and the 1.5-R2-24 (the most superior residue from TSA by hydrothermal method). Incidentally, the spectra of other residues (i.e., 3.0-R2-48 from TSA hydrothermal and 1.4-F1, 1.0-F1, 1.0-F2 and 1.0-F3 from TSA fusion) exhibit significantly sharper band, at 1473.14 cm^{-1} [47]. This is indicative of higher impurities in these residues as well as low *CEC* and hence low grade of zeolites. On the contrary, weaker bands at this wave number in the hydrothermally synthesized residues, 1.5-R2-24 indicates it's less content of impurities, higher *CEC* and hence higher grade of zeolites. Thus 1.5-R2-24 conforms very closely RZP, zeolite 4A, in terms of the *CEC* and grade. This gets verified by identifying the band at 859.8 cm^{-1}, which is also an indication of impurity in the zeolite framework in the form of Si-C-H or Al-C-H bonds. Incidentally, such bands are also present in spectra of fusion residues and not in those of the RZP zeolite and 1.5-R2-24. As far as incorporation of hydroxyl functional group (absent in the RFA) in the form of bonds like [Si-OH-Al] is concerned, a wide variation can be noticed at wave number 3466.3 cm^{-1}. In this context, as compared to other spectra, the fusion residue, 1.0-F3 is observed to have highly hydroxylated framework structure (i.e., Si-OH-Al) and shows development of Bronsted acidic sites [44]. However, the said residue is highly contaminated at 2974 cm^{-1} (i.e., C-H bond, as an impurity) and hence is inferior to the RZP and hydrothermally synthesized zeolite in the residue, 1.5-R2-24, where this band is absent. As an additional advantage, all the bands at 1650.3 (present in most of the spectra) elucidate bending vibration of H-OH bonds and are indicative of adsorbed water on the surface of the zeolite frame work [48]. Apart from these bands, bands at 1091.5 cm^{-1} (in the RFA) and 1008 cm^{-1} in the residues and the RZP provide information of the asymmetric stretching vibration in the Si-O-Al bond. As such a shift of these bands by 83.5 cm^{-1} is a representative of significant coordination of SiO_4 and AlO_4 tetrahedra in the residues and its zeolitic frameworks, as compared to RFA. Further, presence of impurities in the framework (i.e. Si-C-H or Al-C-H bands) of zeolites in the fusion residues makes them inferior residues as compared to that of the hydrothermal residue, 1.5-R2-24 and the RZP (the zeolite 4A).

6.8 Conclusions

Based on the findings presented in this chapter, it can be concluded that the three-step activation of the hopper ash with NaOH results in minor variation in pH and reduction in electrical conductivity of the supernatant. Such activation is also responsible for reduction in Si and Al contents of the supernatant, obtained after recycled treatments. Hence, the final grade of the fly ash zeolites gets improved (with high *CEC* and specific surface area, enhanced specific gravity, nano-sized fine particles of zeolites and micro-sized new pores). The three-step activation of the fly ash by adopting the fusion technique has also been found to be effective for zeolitization of the fly ash. However, the presence of impurities in the ash residues obtained from the fusion process makes them inferior as compared to those obtained from the hydrothermal treatment.

References

1. JCPDS: Joint committee on powder diffraction standards, Philadelphia-19103 (1994)
2. Nugteren, H.W., Moreno, N., Sebastia, E., Querol, X.: Determination of the available Si and Al from coal fly ashes under alkaline conditions with the aim of synthesizing zeolites products.

In: International Ash Utilization Symposium, Centre for Applied Energy Research, University of Kentuchy, Paper No. 71 (2001)
3. ASTM C618-08: Standard specification for coal fly ash and raw or calcined pozzolan for use in concrete, Annual Book of ASTM Standard, 04.08.19, ASTM, Philadelphia, USA
4. Belardi, G., Massimila, S., Pig, A.L.: Crystallization of K-L and K-W zeolites from fly ash. Res. Conserv. Rec. **24**, 167–181 (1998)
5. Singh, D.N., Jha, B., Srinivas, K.: Determination of crystalinity of alkali activated fly ash by XRD and FT-IR studies. In: Yang, Q., et al. (eds.) Constitutive Modeling of Geomaterials, Springer Series in Geomechanics and Geoengineering, pp. 477–481 (2013)
6. Wałek, T.T., Saito, F., Zhang, Q.: The effect of low solid/liquid ratio on hydrothermal synthesis of zeolites from fly ash. Fuel **87**, 194–199 (2008)
7. Kolay, P.K., Singh, D.N., Murti, M.V.R.: Synthesis of zeolites from lagoon ash. Fuel **80**, 739–745 (2001)
8. Rayalu, S., Meshram, S.U., Hasan, M.Z.: Highly crystalline faujasitic zeolites from fly ash. J. Hazard. Mater. **B77**, 123–131 (2000)
9. Wang, D.J., Tang, Y., Dong, A.G., Zhang, Y.H., Wang, Y.J.: Hollow cancrinite zeolite spheres in situ transformed from fly ash cenosphere. Chin. Chem. Lett. **14**(12), 1299–1302 (2003)
10. Jha, B., Singh, D.N.: A review on synthesis, characterization and industrial application of fly ash zeolites. J. Mater. Edu. **33**(1–2), 65–132 (2011)
11. Kolay, P.K., Singh, D.N.: Physical, chemical, mineralogical and thermal properties of cenospheres from a ash lagoon. Cem. Conc. Res. **31**(4), 539–542 (2001)
12. Kolay, P.K., Singh, D.N.: Effect of zeolitization on physic-chemico-mineralogical and geotechnical properties of the lagoon ash. Canad. Geotech. J. **38**(5), 1105–1112 (2001)
13. Criado, M., Fernandez-Jimenez, A., Palomo, A.: Alkali activation of fly ash: Effect of the SiO_2-Na_2O ratio, Part I: FT-IR study. Micropor. Mesopor. Mater. **10**, 180–191 (2007)
14. Larry, R.: Examples of image analysis using ImageJ. Department of Biology, Millersville University, Millersville, PA 17551 (2007)
15. Hollman, G.G., Steenbruggen, G., Janssen, J.M.: A two-step process for the synthesis of zeolites from coal fly ash. Fuel **78**, 1225–1230 (1999)
16. Jha, B., Singh, D.N.: Zeolitization characteristics of a fly ash from wet- and dry- disposal systems. Acta. Geotech. Slov. **2**, 63–71 (2012)
17. Jha, B., Singh, D.N.: A three step process for purification of fly ash zeolites by hydrothermal treatment. App. Clay. Sci. **90**, 122–129 (2014)
18. Jha, B., Singh, D.N.: Formation of meso- and micro-pores in fly-ash zeolites using three step activation. Acta. Geotech. Slov. **1**, 63–69 (2014)
19. Jha, B., Singh, D.N.: Quantification of transitions occurring in a hydrothermally activated fly ash. Mater. Perform. Charact. **3**(1), 239–254 (2014)
20. Rungsuk, D., Apiratikul, R., Pavarajarn, V., Pavasant, P.: Zeolite synthesis from coal-fired power plant by fusion method. In: The 2nd International conference on sustainable energy and environment, 21–23 Nov, Bangkok, Thailand (2006)
21. Grutzeck, M.W., Siemer, D.D.: Zeolites synthesized from class F fly ash and sodium aluminate slurry. J. Am. Ceramic. Soc. **80**(9), 2449–2453 (1997)
22. Rayalu, S.S., Udhoji, J.S., Meshram, S.U., Naidu, R.R., Devotta, S.: Estimation of crystallinity in flyash-based zeolite-A using XRD and IR spectroscopy. Curr. Sci. **89**(12), 2147–2151 (2005)
23. Jacob, P.A., Mortier, W.J.: An attempt to rationalize stretching frequencies of lattice hydroxyl groups in hydrogen zeolites. Zeolite. **2**, 226–230 (1982)
24. Sun, D.M.: An analysis of the structure of 13X molecular sieve in ion exchange. Vacuum **44** (2), 75–78 (1993)
25. Sun, D.M.: Dependence of X zeolite adsorption properties on electronegativity and vibration frequency. Vacuum **45**(12), 1175–1179 (1994)

26. Coates, J.: Interpretation of infrared spectra, a practical approach. In: Meyers, R.A. (ed.) Encycloedia of Analytical Chemistry, pp. 10815–10837. John Wiley & Sons Ltd., Chichester (2000)
27. Li, G.: FT-IR studies of zeolite materials: characterization and environmental applications. Theses and Dissertations, University of Iowa (2005). http://ir.uiowa.edu/etd/96
28. Inada, M., Eguchi, Y., Enomoto, N., Hojo, J.: Synthesis of zeolite from coal fly ashes with different silica-alumina composition. Fuel **84**, 299–304 (2005)
29. Brouwers, H.J.H., Vaneijk, R.J.: Fly ash reactivity: extension and application of a shrinking core model and thermodynamic approach. J. Mater. Sci. **37**, 2129–2141 (2002)
30. Wilkin, R.T., Barnes, H.L.: Solubility and stability of zeolites in aqueous solutions: I. Analcime, Na-, and K-Clinoptilolite. Am. Miner. **83**, 746–761 (1998)
31. Thuadaij, P., Pimraksa, K., Nuntiya, A.: Synthesis of high cation exchange capacity faujasite from high calcium fly ash. Aus. J. Basic. App. Sci. **6**(10), 194–208 (2012)
32. Kim, W., Seung, H., Ahn, B.J.: Synthesis of Na-P1 zeolite from coal fly ash. J. Indus. Eng. Chem. **3**(3), 185–190 (1997)
33. Kolay, P.K., Singh, D.N.: Characterization of alkali activated lagoon ash and its application for heavy metal retention. Fuel **8**, 483–489 (2002)
34. Scott, M.A., Kathleen, A.C., Dutta, P.K.: Handbook of Zeolite Science and Technology. CRC Press, ISBN: 0-8247-4020-3, New York (2003)
35. Tanaka, H., Eguchi, H., Fujimoto, S., Hino, R.: Two-step process for synthesis of a single phase Na-A zeolite from coal fly ash by dialysis. Fuel **85**, 1329–1334 (2006)
36. Querol, X., Moreno, N., Alastuey, A., Juan, R., Andres, J.M., Lopez-Soler, A., Ayora, C., Medinaceli, A., Valero, A.: Synthesis of high ion exchange zeolites from coal fly ash. Geologica. Acta. **5**(1), 49–57 (2007)
37. Querol, X., Moreno, N., Uman, J.C., Alastuey, A., Hernandez, E., Lopez-Soler, A., Plana, F.: Synthesis of zeolites from coal fly ash: an overview. Inter. J. Coal. Geol. **50**, 413–423 (2002)
38. Querol, X., Alastuey, A., Moreno, N., Alvarez, A.E., García-Sánchez, A., Cam, J., Ayora, C., Simon, M.: Immobilization of heavy metals in polluted soils by the addition of zeolitic material synthesized from coal fly ash. Chemosphere **62**, 171–180 (2006)
39. Querol, X., Moreno, N., Uman, J.C., Juan, R., Hernandez, S., Fernandez, P.C., Ayora, C., Janssen, M., Garcıa, J.M.J., Linares, S.A., Cazorla, A.D.: Application of zeolitic material synthesised from fly ash to the decontamination of waste water and flue gas. J. Chem. Technol. Biotechnol. **77**, 292–298 (2002)
40. Santen, R.A.V., Man, A.J.M.D., Kramer, G.J.: Chemical bonding in zeolites, microporous solids: synthesis, structure, and reactivity. Kluwer Academic Publishers, Netherlands (1992)
41. ASAP 2020: Accelerated surface area and porosimetry system. Operator's Manual V3.0, Micromeritics Instrument Corporation, 2004–2006
42. Neimark, A.V., Sing, K.S.W., Thommes, M.: Surface area and porosity. In: Handbook of Heterogeneous Catalysis, Wiley (2008)
43. Franus, W.: Characterization of X type zeolite prepared from coal fly ash. Pol. Environ. Stud. **21**(2), 337–343 (2012)
44. Ojha, K., Pradhan, N.C., Samanta, A.N.: Zeolite from fly ash synthesis and characterization. Bull. Mater. Sci. **27**(6), 555–564 (2004)
45. Williams, D.D., Grand, J.A., Miller, R.R.: The reactions of molten sodium hydroxide with various metals. J. Am. Chem. Soc. **78**(20), 5150–5155 (1956)
46. Van Reeuwijk, L.P.: High-temperature phases of zeolites of the natrolite group. Am. Miner. **57**, 499–510 (1972)
47. Jha, B., Singh, D.N.: Synthesis of higher grade fly ash zeolites X from fly ash via three-step fusion. Mater. Perform. Charact. **2**(1), 1–12 (2013)
48. Jha, B., Koshy, N., Singh, D.N.: Establishing two-stage interactions between fly ash and NaOH by X-ray and infrared analyses. Front. Environ. Sci. Eng. **9**(2), 216–221 (2015)

Chapter 7
Applications of Fly Ash Zeolites: Case Studies

Abstract As a word of mouth, zeolites were first employed for industrial applications by the sugar industries, way back in 1896, which recognized their high adsorptive capacity. In the late 60s, the catalytic properties of zeolites for petrochemical processes, especially the fluid catalytic cracking (FCC) process, were discovered. Currently, around 300,000 tons of synthetic zeolites are being used annually, in catalytic and adsorptive applications. In addition, 250,000 tons of natural zeolites (mainly Clinoptilolite) are being consumed, per annum, as additives in the construction materials and paper industries, waste water treatment, soil improvement, animal feed additive and as cat litters. The potential of industrial applications of zeolitic materials vary as a function of the prevalent zeolite type. The X type zeolite has a large pore size (7.3 Å) and a high *CEC* (500 meq./100 g), and these properties make them an interesting zeolites for separation process which led to it name as a "molecular sieve" and a high cation exchange material, as compared to Hydroxy-sodalite having less pore size (2.3 Å). The zeolites with high Al/Si have been opined to account for the high *CEC* (viz., Na-P1, 4A, X, KM, Chabazite, Herschelite, and Faujasite). Owing to their high *CEC*, these zeolites have a high potential for their application in waste water decontamination, the removal of heavy metals and ammonium from solutions. The potential applications of synthesized ash zeolites are presented in the following sections.

Keywords Fly ash · Zeolites · Application · Absorbant · Ion exchanger · Heavy metals

7.1 Detergent Builder

The conventional detergent builders [viz., sodium tripolyphosphate (STPP) and tetra sodium pyrophosphate (TSPP)], have been identified as environmentally unsafe as they cause excessive deposition of phosphates in the water bodies and result in eutrophication. This is a process which causes excessive growth of algae in water bodies, which once died and decomposed, can cause growth of organic

B. Jha and D.N. Singh, *Fly Ash Zeolites*, Advanced Structured Materials 78,
DOI 10.1007/978-981-10-1404-8_7

matters, which is responsible for depletion of the oxygen level. This can be the main cause of the death of the aquatic lives and biodiversity in the ecosystem.

With this in view, the focus of the researchers has shifted towards the search for the substitutes to such detergent builders [1]. Based on the superiority of zeolite-A with reference to the essential properties for a suitable detergent (viz., ability to lower the hardness of water, insolubility in water and less environmental risk, it has been reported to be dominating among common substitutes to the phosphate builders for laundry detergents with a dose of 25 % by weight of the final product) [2, 3].

Moreover, the cations (viz., Ca^{2+} and Mg^{2+}) present in hard water, are exchanged by zeolite-A by virtue of its "ion sieving" characteristics and their dispersion in the aqueous phase. It has been found that sodium zeolite-A is effective as a water softener agent by ion exchange of the Na^{+} of the zeolite for the Ca^{2+} of the input water stream, and hence it has been used as filler in detergents. In fact, host cations like $Na^{+}/K^{+}/Ca^{2+}$ of the zeolite, present in their hydrated form, are loosely bonded to the aluminum framework resulting in a high degree of mobility. With this in view, they can be removed from the zeolite by exchange with other cations available in water. In reality, the ion exchange behavior of zeolites depends on a number of factors (viz., cation change, cation size, its concentration in solution, temperature of the solution, the solvent type, the structural characteristics of particular zeolite in question, the charge density of zeolite and the specific location of cationic site). In addition, there are certain zeolites where cations could be in an entrapped and inaccessible position for exchange. For example, the sodium cations in sodalite cage, and in the 6-ring portion of zeolite Y, are not easily accessible to ammonium cations. Furthermore, the replacement of sodium ion from zeolite 4A has been found to be dependent on its ion selectivity and degree of affinity for different cations as represented by the sequence; Fe > Al > Cu > Pb > Cd = Tl > Zn > Mn > Ca = Sr > Mg [4]. It can be noticed that the sodium cation of zeolite 4A, has less affinity for calcium and magnesium (i.e., from the detergent point of view) as compared with the other ions, which depend on the factors listed above. In line with this, zeolites A, P and AX have been investigated for their application as detergent builders for the replacement of STPP (i.e., Sodium tripolyphosphate) [5].

Hui and Chao [6] have synthesized the single phase, chamfered-edge zeolite 4A samples in pure form with a high crystallinity, synthesized by step-change of temperature and time during hydrothermal treatment of coal fly ash. The calcium binding capacity of zeolite 4A and its commercial detergent grade (Valfor 100) has been tested for its application as a detergent builder. The authors have clarified that the synthesized zeolite 4A behaves similar to the commercial one, as far as removal of calcium ions during the washing cycle is concerned, as depicted in Fig. 7.1. It has also been reported that the zeolite 4A samples leached the elements Sb, As, Se and Ti to the same extent as their commercial counterparts. The toxicological effects of the ash zeolite 4A have been reported to be as good as that of the commercial grade in removing heavy metal ions from the waste water.

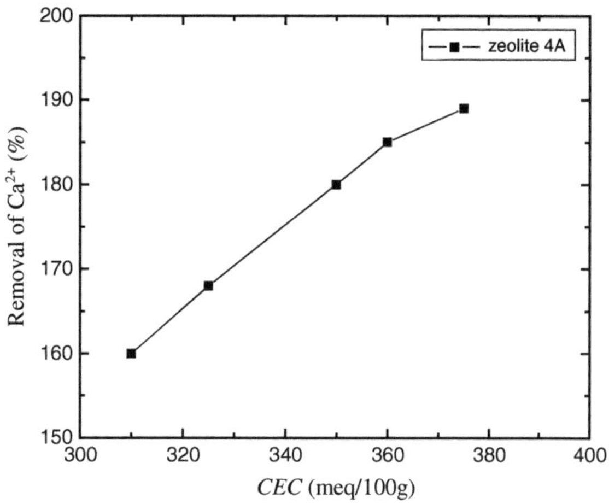

Fig. 7.1 Removal efficiency of calcium ion from water during washing by zeolite 4A bearing detergent

7.2 Heavy Metal Uptake

High ion exchangers like zeolite P, Na-P1, 4A, X, KM, Chabazite, Herschelite and Faujasite with high *CEC* values up to 500 meq./100 g and good sorption potential, have been found to be quite useful for decontamination of sludge, industrial effluents and other waste water by removal of their heavy metals like: Pb^{2+}, Cd^{2+}, Cu^{2+} and Zn^{2+}. Their application for removing heavy metals and ammonia from the sludge is well established [4, 7, 8].

It has been reported that the equilibrium data can be found to fit well in the Langmuir model of sorption [9, 10]. The rate constants have been derived by the relationship shown in Eq. (7.1).

$$\log(q_c - q) = \log q_c - (k/2.303) \cdot T \tag{7.1}$$

where, q_c and q (mg/l) are the amount of metal sorbed at equilibrium and at any time T (min), respectively, and k (min^{-1}) is the rate constant for sorption of metal ions.

The exchange process in zeolite is governed by two mechanisms namely mass transfer of the sorbate for exchange on the surface of zeolite and diffusion within the pores of the zeolites for exchange with the mobile ion corresponding to the lower and higher concentrations of the zeolite dose, respectively. Waste water and flue gas treatment can be performed by application of zeolites A, Y and a blend of zeolites 4A-X. The sludge and/or waste water which has initial pH (≈6.5–7.6) and different concentrations of arsenic (As), ammonia (NH_4^+) and copper (Cu), can be treated by application of proper zeolite dose (g/l). The removal efficiency of the zeolites for these ions is depicted in Fig. 7.2a, b [9, 11].

Shih and Chang [12] have studied the process of conversion of fly ash into zeolites suitable for ion-exchange applications for waste water treatment and

Fig. 7.2 **a** Removal of arsenic and ammonia from waste water by zeolite application. **b** Removal of copper and ammonia from waste water by zeolite Y

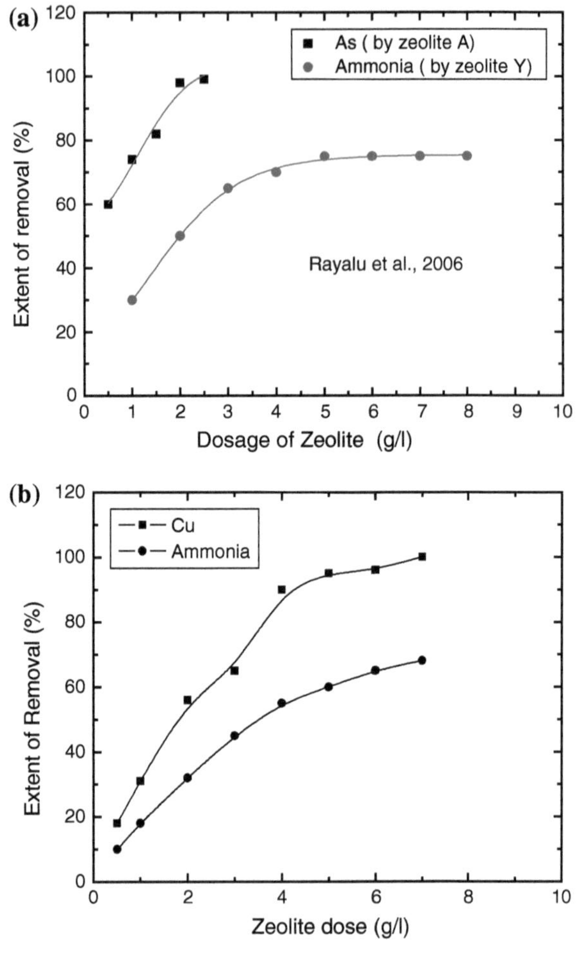

radioactive waste immobilization. It has been reported that Faujasite (i.e., $Na_2O{\cdot}Al_2O_3{\cdot}4.5SiO_2{\cdot}9{\cdot}H_2O$) and zeolite P (i.e., $Na_2O{\cdot}Al_2O_3{\cdot}2SiO_2{\cdot}4.5H_2O$) can be synthesized by treatment of fly ash at 38 and 80 °C, respectively, after 3 days. It has been observed that less stable Faujasite and zeolite A can be formed by maintaining the Si/Al ratio to less than unity during the synthesis process by addition of $Al(OH)_3$ in the reaction mixture of fly ash and NaOH. These researchers have concluded that higher Si/Al molar ratio and temperature are conducive to form zeolite P. It has been demonstrated that the zeolite A and Faujasite can result into more exchange of NH_4^+ than Cs^+ ion because of its larger ionic size.

Ma et al. [13] have studied the cation exchange properties of zeolites Na-P type synthesized from fly ash under hydrothermal conditions by employing 2.8 and 5 M NaOH concentrations at 100 °C. It has been demonstrated that the zeolites produced have fibrous morphology, well defined crystal faces with specific surface area ranging from 28.5 to 41.1 m^2/g; corresponding to the increase in NaOH

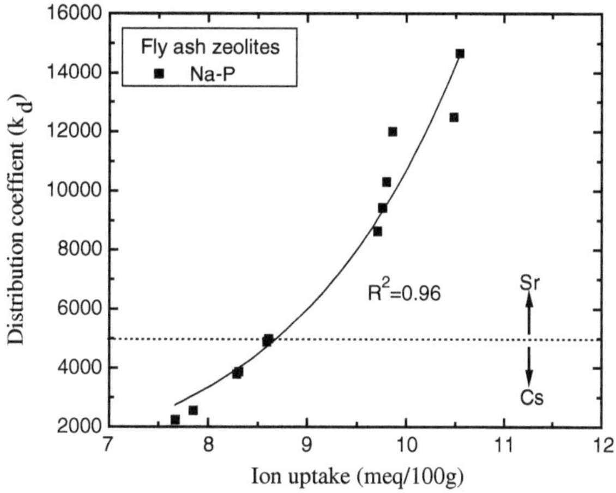

Fig. 7.3 Variation of distribution coefficient with ion uptake capacity of the synthesized zeolites

concentration from 2.8 to 5 M, respectively. Based on these attributes, the synthesized zeolites can show significant selectivity for the uptake of Cs and Sr cations as depicted in Fig. 7.3.

Steenbruggen and Hollman [14] have demonstrated that fly ash can be converted up to 45 % into zeolitic materials. They have opined that different types of zeolites e.g. Na-P1, KG and ZK19 can be synthesized by varying the experimental conditions so as to increase their *CEC* from 2 to 240 meq./100 g. It has been confirmed that the ion selectivity series for zeolite Na-P1 can follow the trend like Ba > Cu > Cd > Zn > Co > Ni. The cation exchange reactions as well as precipitation of hydroxides can play important role in the immobilization of heavy metals. The results of the sorption experiments for the zeolitic products have further highlighted their application in environmental pollution controlas an immobilizer of pollutants.

Kolay and Singh [15] have investigated the alkali activated lagoon fly ash for its heavy metal retention capacity. It has been opined that a 12 h interaction of activated ash samples which contains zeolite Na-P1, with 0.0005 M $Pb(NO_3)_2$ can result into 100 % removal of Pb^{2+} from similar type of industrial sludge which has initial pH from 2 to 8 and ion concentration of the order of 3885 ppm.

Querol et al. [16, 17] have reported that the type of industrial application of the synthesized end product is dependent on the type of the zeolite grade crystallized. It has been opined that zeolite X can be an interesting material to be applied as a molecular sieve and high cation exchanger which can be attributed to its large pore size (≈7.3 Å) and high *CEC* (≈500 meq./100 g). However, the small pore size of Hydroxy-sodalite (≈2.3 Å) has been reported to account for its low potential application. The uptake of heavy metals from polluted waste water has been an established application of zeolites, Na-P1 and 4A type by employing their doses in different ranges: 5–30 g/l and 10–25 g/l, respectively. In addition, it has been demonstrated that a blend of pure zeolites 4A and X, can get formed from fly ash, by employing silica extraction techniques, which can result in the *CEC* of the activated product up to 470 meq./100 g.

Scott et al. [18] have established that the synthesized fly ash zeolites can be a suitable material for the removal of lead ions from aqueous solution within 24 h of application. They have reported that type of zeolites (viz., zeolites X, Na-P1, Sodalite and Phillipsite) synthesized in the hydrothermal method can depend on the treatment time, concentration of NaOH solution and the type of fly ash employed in the process. It has been concluded that the type of zeolites and their morphology can influence the rate of lead ion removal.

Derkowski et al. [19] have studied the properties and potential applications of zeolitic materials produced from fly ash using simple method of synthesis. It has been demonstrated that zeolitic materials and their framework types like Na-X (i.e., FAU type), Na-P1 (i.e., GIS type) and Sodalite (i.e., SOD type) can be synthesized by employing three different mixtures of fly ash, NaOH and NaCl under atmospheric pressure and at temperatures below 110 °C. It has been concluded that the *CEC* and the uptake capacity of heavy metal ions of synthesized zeolites, are of the order of 5.5–239 meq./100 g and 40 mg/g, respectively. They have confirmed the general applications of synthesized zeolitic materials for removal of heavy metal ions from waste water, adsorption or refining of used lubricating oils and separation of molecules of different sizes because of their high *CEC* and ion-sieving properties. It has been opined that zeolite Na-X can be a substitute of commercially available zeolite 13X, for its application as a gas adsorbent. It has been confirmed that the synthesized zeolites, 4A and X type, can be suitable materials for water purification and soil remediation, whereas zeolite Na-P1 can be useful for adsorption of metal ions from sludge and soil, oil refining, geosynthetic liners and ground water remediation.

Somerset et al. [20] have studied the process of the hydrothermal conversion of fly ash into zeolites (viz., Faujasite, Sodalite and zeolite A) and their application for removal of mercury and lead ions from wastewater. It has also been reported that a dose of zeolites of the order of 5–20 g/l can result the removal of Pb and Hg by 95 and 30 %, respectively, from the acid mine drainage (AMD) at an average pH of 4.5.

Jha et al. [21] have studied the process of zeolite formation from coal fly ash and heavy metal ion removal characteristics of zeolite X in multi-metal system. These authors have reported that the active carbon present in the fly ash can enhance the formation of zeolite Na-X. It has been opined that zeolite, Na-A can be initially crystallized by employing hydrothermal activation of fused coal fly ash (CFA), which can later get transformed to zeolite Na-X corresponding to the increase of NaOH concentration and/or reaction time. It has been confirmed that the synthesized zeolites, Na-X, can be applied for removal of 68 % of Ni^{2+}, Cd^{2+}, Cu^{2+} and Pb^{2+} from waste water, contaminated with multi metal ions, based on its *CEC* of the order of 3.2 meq./g as compared to 4.7 meq./g of its commercial grade.

Qiu et al. [22] have demonstrated a technique for removal of Lead, Copper, Nickel, Cobalt, and Zinc from water by application of a Cancrinite type zeolite synthesized from fly ash, by employing molten salt method. The degree of heavy metal adsorption has been reported to vary with the type of ion exchange mechanism, Si/Al ratio and pore size of Cancrinite and the pH of the solution. In contrast to this, it has been opined that a larger specific area (≈278.9 m^2/g) can augment the diffusion process of heavy metal cations into the pores and channels of a zeolitic framework.

7.3 As an Adsorbent

The pores on the external surface of zeolites provide access into their internal channels for molecules of smaller size, shape and characteristics inductive to such permeation followed by diffusion. This may be the reason that zeolites possess high adsorption/desorption characteristics without any effect on their lattice structure, which enables them to store and release solar energy. Based on the literature, it has been opined that the adsorption is a process related to the decrease in the freedom of molecule, its entropy and hence the release of heat of adsorption (i.e., exothermic reaction), whereas, it's reverse process (i.e., desorption), can favor storage of heat (i.e., endothermic reaction). The important parameter is the type of adsorbent (viz., aluminum oxide as alumina gel or activated alumina, charcoal or activated carbon, silicagel, a porous material and crystalline zeolites, alumino silicates, with porous structure usually in pellet or spherical form, which can adsorb large quantity of water. In fact, zeolites (viz., 13X, 4A and 5A) can be substitutes to the expensive silica gel for this purpose, which can be attributed to their better performance in case of variation in relative humidity and temperature. Another important parameter is the heat of adsorption, which can depend, slightly, on the zeolite structure. For example, 1000 kg of zeolite 13X can result in the storage of 50,000 kJ of energy per minute for solar powered heaters [23]. Based on this principle of adsorption-desorption, zeolites are also being used in air conditioning systems to reduce moisture in the air and maintain effective evaporative cooling.

In addition, they have potential applications as an additive in soils, for heavy metal adsorption [24] and as carrier in pesticides, herbicides and fungicides. Based on such adsorbing characteristics, zeolites have also been effective as additive in animal feeds to act as a detoxing agent, e.g., Clinoptilolite. Zeolites exhibit possibilities of applications in animal sheds for adsorbing hydrogen sulphides and subsequent odour control in the surrounding environment.

Querol et al. [25] have presented an overview on the methodologies for zeolite synthesis from the fly ash. The authors have detailed the conventional alkaline conversion processes, with special emphasis on the experimental conditions to obtain high cation exchange capacity (*CEC*) zeolites. They have reported that zeolitic products having *CEC* up to 300 meq./100 g, can be obtained from high-glass fly ash by direct conversion and the main application of this material is the uptake of heavy metals and ammonium ion from polluted water. It has been clarified that some of the zeolites synthesized, are useful as molecular sieves to absorb water molecules from gas streams or to trap SO_2 and NH_3 from low water gaseous emissions based on their pores and molecular sizes as depicted in Fig. 7.4.

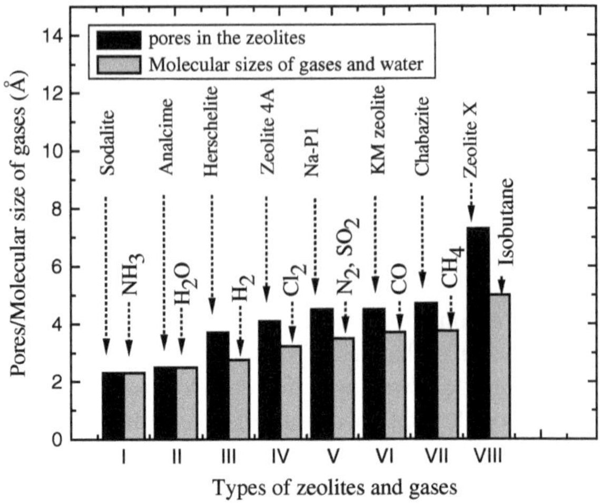

Fig. 7.4 Pore size of zeolites and the molecular size of various gases suitable for adsorption

Chunfeng et al. [26, 27] have demonstrated the application of zeolite A and X in removal of dye and methylene blue from waste water as an alternative to activated carbon and commercial grade of zeolites. As such, zeolites have been used for catalytic cracking, oxygen production, as dehydrating agent and for isomerization processes in the industries. Zeolite Y and ZSM-5 have been confirmed to be employed for Fluid Catalytic Cracking, FCC [28]. In fact, FCC is the process employed by the petroleum refineries to convert hydrocarbons (i.e., high-boiling, and high-molecular weight) of petroleum crude oils to their lighter fractions (viz., gasoline and other products with better octane numbers) by catalytic cracking [29].

7.4 Separation of Gases

The adsorption of water, hydro carbons and other liquid molecules from the gas streams can be carried by employing zeolites absorption characteristics. Zeolite 3A, 4A, 5A and 13X have been characterized for their application as adsorbents and desiccators [30]. These zeolites have also been found to have potential applications as molecular sieves in gas purification technology [31].

Flue gas treatment can occur by application of zeolite X (pores 7.3 Å), Na-P1, Na-Chabazite, Philipsite-KM, Linde F, Analcime, Sodalite and blend of zeolites 4A and X, which have been investigated to have different *CEC* and adsorption capacity for separation of CO_2, SO_2, H_2O and NH_3 from low water (i.e., the lowest level of water in a body of water such as river, lake, reservoir and well) gaseous emissions which can be attributed to their surface pore sizes as depicted in Fig. 7.5 [25, 32].

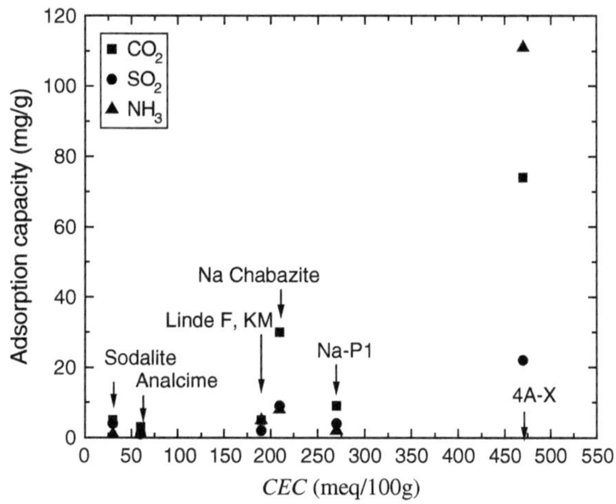

Fig. 7.5 Effect of application of various zeolites in separation of flue gases based on their cation exchange capacity

7.5 Partial Replacement of Cement

Canpolat et al. [33] have suggested the usage of zeolite, coal bottom ash and fly ash as replacement materials for production of the cement. It has been opined that the inclusion of zeolites, up to 15 %, results in an increase in the compressive strength of the cement, at early ages, whereas a decrease in compressive strength is observed when zeolites are used in combination with the fly ash. However, the setting time is found to decrease due to the replacement of clinker by the zeolite. This has been attributed to presence of adsorbed water (the excess water remained unused after the hydration process in the cement) on the zeolite surface.

7.6 Radioactive Waste Treatment

Mimura et al. [34] have studied the alkali hydrothermal synthesis of zeolites from coal fly ash and their uptake properties of Cesium ion. They have synthesized zeolites KH ($K_2Al_2Si_{14}O_{12}{\cdot}nH_2O$) by employing KOH activation and optimized the synthesis process to 1 M KOH concentration, temperature at 160° C, and time of 3 days and L/S of 15 ml/g. It has been opined that a distribution coefficient as large as 10^4 cm^3/g can be achieved by employing zeolite dose for 2 h, corresponding to the uptake of the radioactive Cs^+ ion. In contrast to this, the synthesized product can also exhibit the uptake of Cs^+ ion as high as 3.34 mmol/g as estimated from Langmuir plots, which has been opined to be around 2–3 times greater than that of the synthetic zeolite P and natural Mordenite.

The authors have established relations between the uptake percentage of Cs^+ ions removed from the solution, R_{Cs} (%), and the distribution coefficient, $K_{d{\cdot}Cs}$ (cm^3/g), which can be defined by Eqs. (7.2) and (7.3), respectively.

$$R_{Cs} = [(A_i - A_t)/A_i] \cdot 100 \quad (7.2)$$

$$K_{d \cdot Cs} = \left[(A_i - A_f)/A_f\right] \cdot V/m \quad (7.3)$$

where A_i, A_t and A_f are the counting rates per unit volume (counts per minute, cpm/cm^3) at the initial stage, at any time t, and at equilibrium respectively; m(g) is mass of the product and V(cm^3) is the volume of the aqueous phase.

Mishra and Tiwari [35] have studied the performance of zeolite 13X, synthesized by employing hydrothermal activation of NaOH for 20 h, and fused fly ash at 600 °C. The authors have demonstrated that the BET surface area of the original fly ash (≈6 m^2/g) gets increased up to 430 m^2/g because of the alkali activation, which can be attributed to the formation of the zeolites at par with their commercial grades. It has been reported that these zeolites, can exhibit selectivity of cations corresponding to the trend as $Cu^{2+} > Co^{2+} > Ni^{2+}$ under different pH and temperature conditions.

7.7 Conclusions

The applications of fly ash zeolites discussed in this chapter are summarized here in brief. Fly ash zeolites A, 4A, P and 4X are suitable for detergent builder and common substitute to phosphate builders. Other types of zeolites A, Y, P, Na-P1, 4A, X, KM, Chabazite, Herschelite and Faujasite are effective for removal of heavy metals and ammonia from waste water. Zeolite A and Faujasite are proven cation exchange materials for removal of NH_4^+ from sewage. Zeolite Na-P, Hydroxy-sodalite, Sodalite, Na-X and Philipsite are also established for uptake of Ni, Cd, Cu, Cs, Pb, Hg and Sr cations from contaminated water, soil and gas as adsorbent. In short, zeolites can act as detoxing agent to decontaminate the environment and hence facilitate its conservation.

References

1. http://ec.europa.eu/environment/water/pollution/phosphates/pdf/phosphates_xsum.pdf. Assessed on 5 March 2011
2. Udhoji, J.S., Bansiwal, A.K., Meshram, S.U., Rayalu, S.S.: Improvement in optical brightness of fly ash zeolite-A for use as detergent builder. J. Scientific. Indus. Res. **64**, 367–371 (2005)
3. VEGOBONDS: Zeolites for detergent; U.S. Patent 2009/0075857 (2009)
4. Moreno, N., Querol, X., Ayora, C., Alastuey, A., Fernandez, P.C., Janssen, J.M.: Potential environmental applications of pure zeolitic material synthesized from fly ash. J. Environ. Eng. **127**(11), 994–1002 (2001)
5. Jha, B., Singh, D.N.: A review on synthesis, characterization and industrial application of flyash zeolites. J. Mater. Edu. **33**(1–2), 65–132 (2011)

6. Hui, K.S., Chao, C.Y.H.: Pure, single phase, high crystalline, chamfered-edge zeolite 4A synthesized from coal fly ash for use as a builder in detergents. J. Hazard. Mater. **B137**, 401–409 (2006)
7. Berkgaut, V., Singer, A.: High capacity cation exchanger by hydrothermal zeolitization of coal fly ash. App. Clay. Sci. **10**, 369–378 (1996)
8. Amrhein, C., Haghnia, G.H., Kim, T.S., Mosher, P.A., Gagajena, R.C., Amanios, T., Torre, D. L.: Synthesis and properties of zeolites from coal fly ash. Environ. Sci. Technol. **30**, 735–741 (1996)
9. Rayalu, S.S., Bansiwal, A.K., Meshram, S.U., Labhsetwar, N., Devotta, S.: Fly ash based zeolite analogues: versatile materials for energy and environment conservation. Catal. Surv. Asia **10**(2), 74–88 (2006)
10. Querol, X., Alastuey, A., Moreno, N., Alvarez, A.E.: Garcı´a-Sa´nchez, A., Cam, J., Ayora, C., Simon, M.: Immobilization of heavy metals in polluted soils by the addition of zeolitic material synthesized from coal fly ash. Chemosphere **62**, 171–180 (2006)
11. Rayalu, S., Meshram, S.U., Hasan, M.Z.: Highly crystalline faujasitic zeolites from fly ash. J. Hazard. Mater. **B77**, 123–131 (2000)
12. Shih, W., Chang, H.: Conversion of fly ash into zeolites for ion-exchange applications. Mater. Lett. **28**, 263–268 (1996)
13. Ma, W., Brown, P.W., Komarneni, S.: Characterization and cation exchange properties of zeolites synthesized from fly ashes. J. Mater. Res. **13**(1), 3–7 (1998)
14. Steenbruggen, G., Hollman, G.G.: The synthesis of zeolites from fly ash and the properties of the zeolite products. J. Geochem. Explor. **62**, 305–309 (1998)
15. Kolay, P.K., Singh, D.N.: Characterization of alkali activated lagoon ash and its application for heavy metal retention. Fuel **8**, 483–489 (2002)
16. Querol, X., Moreno, N., Uman, J.C., Alastuey, A., Hernandez, E., Lopez-Soler, A., Plana, F.: Synthesis of zeolites from coal fly ash: an overview. Inter. J. Coal. Geol. **50**, 413–423 (2002)
17. Querol, X., Moreno, N., Alastuey, A., Juan, R., Andres, J.M., Lopez-Soler, A., Ayora, C., Medinaceli, A., Valero, A.: Synthesis of high ion exchange zeolites from coal fly ash. Geologica. Acta. **5**(1), 49–57 (2007)
18. Scott, M.A., Kathleen, A.C., Dutta, P.K.: Handbook of zeolite science and technology. CRC Press, ISBN: 0-8247-4020-3, NewYork (2003)
19. Derkowski, A., Franus, W., Waniak-Nowicka, H., Czimerova, A.: Textural properties vs. *CEC* and EGME retention of Na-X zeolite prepared from fly ash at room temperature. Inter. J. Miner. Process. **82**, 57–68 (2006)
20. Somerset, V., Petrik, L., Iwuoha, E.: Alkaline hydrothermal conversion of fly ash precipitates into zeolites: The removal of mercury and lead ions from wastewater. J. Environ. Manage. **87**(1), 125–131 (2008)
21. Jha, V.K., Nagae, M., Matsuda, M., Miyake, M.: Zeolite formation from coal fly ash and heavy metal ion removal characteristics of thus obtained zeolite X in multi-metal system. J. Environ. Manage. **90**, 2507–2514 (2009)
22. Qiu, W., Zheng, Y.: Removal of lead, copper, nickel, cobalt, and zinc from water by a cancrinite-type zeolite synthesized from fly ash. Chem. Eng. J. **145**, 483–488 (2009)
23. Gopal, R., Hollebone, B.R., Langford, C.H., Shigeishi, R.A.: The rate of solar energy storage and retrieval in a zeolite-water system. Solar. Energ. **28**(5), 421–424 (1982)
24. Singh, D.N., Kolay, P.K.: Simulation of ash water interaction and its influence on ash characteristics. Prog. Energ. Combus. Sci. **28**, 267–299 (2002)
25. Querol, X., Moreno, N., Uman, J.C., Juan, R., Hernandez, S., Fernandez, P.C., Ayora, C., Janssen, M., Garcıa, J.M.J., Linares, S.A., Cazorla, A.D.: Application of zeolitic material synthesised from fly ash to the decontamination of waste water and flue gas. J. Chem. Technol. Biotechnol. **77**, 292–298 (2002)
26. Chunfeng, W., Jiansheng, L., Lianjun, W., Xiuyun, S., Jiajia, H.: Adsorption of dye from wastewater by zeolites synthesized from fly ash: kinetic and equilibrium studies. Chin. J. Chem. Eng. **17**(3), 513–521 (2009)

27. Chunfeng, W., Jiansheng, L., Xia, S., Lianjun, W., Xiuyun, S.: Evaluation of zeolites synthesized from fly ash as potential adsorbents for wastewater containing heavy metals. J. Environ. Sci. **21**, 127–136 (2009)
28. Monographs, I.A.R.C.: Zeolites other than Errionite. **68**, 307–333 (1996)
29. http://en.wikipedia.org/wiki/Fluid_catalytic_cracking. Accessed on 6 March 2011
30. Font, O., Moreno, N., Diez, S., Querol, X., Lopez-Soler, A., Coca, P., Garcia, P.F.: Differential behaviour of combustion and gasification fly ash from Puertollano Power Plants (Spain) for the synthesis of zeolites and silica extraction. J. Hazard. Mater. **166**, 94–102 (2009)
31. Fukui, K., Katoh, M., Yamamoto, T., Yoshida, H.: Utilization of NaCl for phillipsite synthesis from fly ash by hydrothermal treatment with microwave heating. Adv. Powder Technol. **20**, 35–40 (2009)
32. http://www.freedictionery.com/low+water. Accessed on 6 March 2011
33. Canpolat, F., Yilmaz, K., Kose, M.M., Sumer, M., Yurdusev, M.A.: Use of zeolite, coal bottom ash and fly ash as replacement materials in cement production. Cem. Conc. Res. **34**, 731–735 (2004)
34. Mimura, H., Yokota, K., Akiba, K., Onodera, Y.: Alkali hydrothermal synthesis of zeolites from coal fly ash and their uptake properties of Cesium ion. J. Nuc. Sci. Technol. **38**(9), 766–772 (2001)
35. Mishra, T., Tiwari, S.K.: Studies on sorption properties of zeolite derived from Indian fly ash. J. Hazard. Mater. **B137**, 299–303 (2006)

Chapter 8
Path Ahead

Abstract The methods of synthesis of fly ash zeolites, adopted by the previous researchers and those established by the authors of this handbook, yield by-products, as well. These by-products (containing alkaline liquid sludge containing considerable concentration of soluble Si and Al ions) require special attention both in terms of their handling, temporary storage and disposal. In this situation, efforts should be made to develop methodologies that would be effective in safe and environmentally friendly solutions to deal with the siliceous and aluminous enriched liquid wastes. One of the ways, to deal with this situation would be to utilize the liquid sludge coming out of the zeolitization process for sequential hydrothermal treatment so as to create a zero effluent process. Accordingly, this chapter deals with some potential areas for further research on the fly ash and alkali interaction.

Keywords Fly ash zeolites · Application · Pore network · Effluent · Pilot plant

8.1 Application of the Zeolites from the TSA

The synthesized fly ash zeolites by resorting to the hydrothermal three step activation (TSA) technique exhibit high cation exchange capacity, as discussed in Chap. 6 [1–8]. Hence, these zeolites should be utilized for various environmental cleanup projects, as absorbent.

8.2 Pore Network in the Zeolites from the TSA

The pore network of the fly ash zeolites synthesized by the TSA has not been explored yet [5]. These investigations are a must to quantify the volumes of pores in such zeolites, by employing highly sophisticated characterization tools like Mercury Intrusion Porosimetry, BET surface area analyzer etc.

B. Jha and D.N. Singh, *Fly Ash Zeolites*, Advanced Structured Materials 78,
DOI 10.1007/978-981-10-1404-8_8

8.3 A Zero-Effluent Technique for Zeolite Synthesis

As described in Chaps. 5 and 6, the laboratory based three step activation (TSA) of raw fly ash (RFA) has been opined to be an effective technique to synthesize high grade cation exchange material (the fly ash zeolites, FAZ) from the fly ash [3]. Based on the limitations of the said technique that it produces liquid by-products (the final effluent), which contains high concentrations of Si, Al and other trace elements (viz., Mn, Ti, Mg, Pb, P etc.) of the fly ash [2], some innovative solutions should be developed for its reuse and finally safe disposal in the environment.

8.3.1 Synthesis of the Special Fly Ash Based Cement

The siliceous and aluminous contents of the effluent from the TSA [6], should be tested for their affinity to the hydraulic lime to form cementitious substances (termed as the special cement). Once experimentations are planned and attempts are made to monitor the physical, chemical, mineralogical and morphological characteristics of such special cement, the inferences would pave some way to a further step ahead towards establishing an innovative fly ash zeolite synthesis technique with zero-emission waste for a safer environment

8.3.2 Synthesis of New Grade of Zeolites from the Final Effluent

The soluble Si and Al, present in the final effluent after the TSA, as completed by following the procedures described in Chap. 5 [3], can be altered if pure SiO_2 and/or Al_2O_3, in different proportions, are added in the effluent externally. A controlled chemical reaction, in the chemically modified effluent, might be helpful to yield improved zeolites, which in turn would facilitate synthesis of comparatively different grades of zeolites as compared to those obtained from the TSA. This endeavor is also expected to have a zero effluent, as an additional advantage of the fly ash zeolite synthesis process.

8.4 A Pilot Plant Based on Hydrothermal Three-Step Activation

Based on high cation exchange capacity of the fly ash zeolites, as discussed in Chap. 6 [3], it would be quite intriguing to synthesize zeolites, in large quantities, at a pilot plant. A suitable lay-out of such a pilot plant should be carefully prepared. This should follow by generating proper facilities to characterize the zeolites so produced.

8.5 Conclusions

It can be realized that there is a huge scope of conducting research in the fields of synthesis, characterization and adequate utilization of fly ash zeolite. Attempts in the direction of zero effluent zeolite synthesis technique, application of the zeolites from the TSA, quantification of pore network in such zeolites would be prudent for the future researchers. In addition, for large scale synthesis of fly ash zeolites, details of the plant level synthesis should also be worked out.

References

1. Jha, B., Padmakumar, G.P., Singh, D.N., Iyer, K.: Synthesis of zeolites by fly ash alkali interaction. In: Proceedings of Indian Geotechnical Conference, Cochi, India, pp. 1089–1092 (2011)
2. Jha, B., Singh, D.N.: A review on synthesis, characterization and industrial application of fly ash zeolites. J. Mater. Edu. **33**(1–2), 65–132 (2011)
3. Jha, B., Singh, D.N.: A three step process for purification of fly ash zeolites by hydrothermal treatment. App. Clay. Sci. **90**, 122–129 (2014)
4. Jha, B., Singh, D.N.: Zeolitization characteristics of a fly ash from wet- and dry-disposal systems. Acta. Geotech. Slov. **2**, 63–71 (2012)
5. Jha, B., Singh, D.N.: Formation of meso- and micro-pores in fly-ash zeolites using three step activation. Acta. Geotech. Slov. **1**, 63–69 (2014)
6. Jha, B., Singh, D.N.: Quantification of transitions occurring in a hydrothermally activated fly ash. Mater. Perform. Charact. **3**(1), 239–254 (2014)
7. Jha, B., Singh, D.N.: Synthesis of higher grade fly ash zeolites X from fly ash via three-step fusion. Mater. Perform. Charact. **2**(1), 1–12 (2013)
8. Jha, B., Koshy, N., Singh, D.N.: Establishing two-stage interactions between fly ash and NaOH by X-ray and infrared analyses. Front. Environ. Sci. Eng. **9**(2), 216–221 (2015)

Index

B. Jha and D.N. Singh, *Fly Ash Zeolites*, Advanced Structured Materials 78,
DOI 10.1007/978-981-10-1404-8

MIX
Papier aus verantwortungsvollen Quellen
Paper from responsible sources
FSC® C105338

If you have any concerns about our products,
you can contact us on
ProductSafety@springernature.com

In case Publisher is established outside the EU,
the EU authorized representative is:
Springer Nature Customer Service Center GmbH
Europaplatz 3, 69115 Heidelberg, Germany

Printed by Libri Plureos GmbH
in Hamburg, Germany